Mathematics in Mind

The monographs and occasional textbooks published in this series tap directly into the kinds of themes, research findings, and general professional activities of the **Fields Cognitive Science Network**, which brings together mathematicians, philosophers, and cognitive scientists to explore the question of the nature of mathematics and how it is learned from various interdisciplinary angles. Themes and concepts to be explored include connections between mathematical modeling and artificial intelligence research, the historical context of any topic involving the emergence of mathematical thinking, interrelationships between mathematical discovery and cultural processes, and the connection between math cognition and symbolism, annotation, and other semiotic processes. All works are peer-reviewed to meet the highest standards of scientific literature.

More information about this series at http://www.springer.com/series/15543

Stacy A. Costa • Marcel Danesi
Dragana Martinovic

Editors

Mathematics (Education) in the Information Age

 Springer

Editors
Stacy A. Costa
Curriculum, Teaching & Learning
& Engineering Education Collaboration
Ontario Institute for Studies in Education
University of Toronto
Toronto, ON, Canada

Marcel Danesi
Department of Anthropology
University of Toronto
Toronto, ON, Canada

Dragana Martinovic
Faculty of Education
University of Windsor
Windsor, ON, Canada

ISSN 2522-5405 ISSN 2522-5413 (electronic)
Mathematics in Mind
ISBN 978-3-030-59179-3 ISBN 978-3-030-59177-9 (eBook)
https://doi.org/10.1007/978-3-030-59177-9

This Springer imprint is published by the registered company Springer Nature Switzerland AG
The registered company address is: Gewerbestrasse 11, 6330 Cham, Switzerland

Preface

Changes in information technologies lead to concomitant changes in social group-think. This was a principle put forth by the Canadian communications theorist Marshall McLuhan. His most famous exemplification of this principle was the invention of the alphabet, around 1000 BCE, which allowed for information to be preserved and thus used over and over, modified, and elaborated. He called this the first true paradigm shift in human consciousness. Although he passed away before the current era, aptly called the Information Age, there is little doubt that he would say that a second paradigm shift, based on computer technologies, has taken place. This bears many important implications for everything we do and how we think about things.

The Information Age impels all of us to become more involved with one another, no matter what language we speak or what culture we come from. This has engendered new perceptions of what knowledge and education are or should be, leading to new ways of learning and researching, such as crowdsourcing and other collaborative modes of interacting. This is certainly true of mathematics. In a relevant book, *Math bytes* (Princeton: Princeton University Press, 2014), Tim Chartier argues how some ideas, like Google's algorithms, are changing how people now view mathematics. So, are the traditional ways of doing and teaching mathematics still viable? This collection of essays deal with this main question from various angles. These look at both the "positives" and the "negatives" and how to reconcile the past with the present in mathematics education.

The opening chapter by Costa, Danesi, and Martinovic provides an overview of how mathematics is practiced and taught in the Information Age. The authors look at crowdsourced mathematics and its implications for mathematics education. Because of technology, mathematicians have a powerful new socially based way to do mathematics, called "massively collaborative mathematics," and this has had significant implications for mathematics education. In Chap. 3, Krpan and Sahmbi make a similar kind of cogent plea for reviving argumentation and reasoning skills in the elementary classroom. In Chap. 9, Gollish makes a plea for restoring "fun" in learning mathematics by blending traditional and new ways of teaching. Amidst all the enthusiasm for new technologies, new mathematical techniques, etc., the "basics" and the "traditional" methods of proof are probably more necessary today than in the past, given that they have receded to the margins of the elementary mathematics classroom. Without these, the development of a deeper knowledge required to grasp mathematics might never crystallize in learners. In Chap. 12, Metz and Davis outline what it means to teach mathematics in the Information Age, constituting a kind of overview of the various pedagogical themes of the book. The psychosocial implications of the technology revolution are thus applied to the mathematics classroom.

Popular culture, social media, and mathematics (education) are the themes of Chaps. 2, 4, and 6. In Chap. 4, Danuser looks at the role of women in STEM subjects and why women continue to be underrepresented, despite all their achievements in these subjects. Using concepts from the field of visual rhetoric, Danuser decodes the Ad Council's "She Can STEM" campaign, which promotes STEM to young women. She shows how the campaign materials aim to subvert the culturally dominant stereotype that science is a masculine endeavor. She also looks at the shortcomings in the campaign. In Chap. 6, Nuessel looks at how mathematics is not only part of modern sports but also intrinsic to understanding them. In this overview, Nuessel derives pedagogical implications for incorporating the simple use of statistics in sports into mathematical pedagogy. Although well known, the relevant statistical techniques in sports are summarized here and related to an overall approach to teaching that involves the use of this sector of popular culture as a platform on which to introduce ideas. The second chapter, by Dan Vilenchik, looks at how to make sense of the massive data that are automatically collected from online platforms such as online social media or e-learning platforms. He puts forward two main methods of finding patterns in the data in an unsupervised manner: clustering and low-dimensional approximation, covering both the theoretical aspects of each one and providing real-data examples. The chapter has implications for how social media are perceived and used.

The topic of diagrams and graphs in mathematics is covered in three chapters. In Chap. 5, Costa examines STEM subjects from the point of view of how information is represented graphically in them and how graphical literacy should not be assumed but taught explicitly. In Chap. 7, Kauffman shows how diagrams link mathematical fields, examining diagrammatic aspects of the vector calculus and then showing how and why they work by relating them to the question of coloring maps and graphs in and out of the plane. He also provides an in-depth treatment of knot theory diagrams, showing how theory is related to non-associative algebras called quandles. In the first part of the chapter, Kaufmann uses a fictional dialogue format to present difficult ideas conversationally, making them easier to follow. Of special interest is Kaufmann's illustration of how diagrammatic principles apply to the classic Four-Color Theorem. In Chap. 11, Kiryushenko explains the power of diagrams in mathematical discovery, using the Existential Graphs of Charles Peirce to make his case. These have become intrinsic to many aspects of set theory today. Kiryushenko also shows how mathematical thinking is a visual Gestalt, rather than a purely linguistic-logical one. The implications for mathematics education are also discussed in a generic way.

In Chap. 8, Danesi looks at the emergence of experimental mathematics as a branch of both mathematics and computer science, tracing its origins and discussing its value to both the practice of mathematics today and the pedagogical implications that "automated mathematics" harbors for the contemporary mathematics classroom. In Chap. 10, Francis et al. deal with the use of curated robotics in the mathematics classroom, thus expanding and applying the idea of incorporating machines in the classroom as devices to enhance learning. They show how a well-structured robotics inquiry allows students to discern critical features of a concept via multiple instantiations of the concept. Finally, in Chap. 13, Logan provides a cogent argument that language and mathematics are intertwined cognitively and representationally. His excursus into the origins of both faculties is persuasive, providing a basis on which to understand how language and mathematics cannot be separated in any real way.

Today, there is great enthusiasm and optimism about technology. However, as in any paradigm shift, there are a number of disadvantages, leading to a consideration of caveats. The chapters in this book, overall, look at these as well from different angles. In his book, *The age of missing information* (New York: Random House, 1992), journalist Bill McKibben issued an overall caveat at the dawn of the Information Age that is worth repeating here, since it is a subtext in various chapters of this book:

> We believe that we live in the 'age of information,' that there has been an information 'explosion,' an information 'revolution.' While in a certain narrow sense that is the case, in many more important ways just the opposite is true. We also live at a moment of deep ignorance, when vital knowledge that humans have always possessed about who we are and where we live seems beyond our reach. An unenlightenment. An age of missing information (p. 9).

This book is part of a series of projects undertaken at the Fields Institute for Research in Mathematical Sciences under the aegis of its *Cognitive Science Network: Empirical Study of Mathematics and How It Is Learned*. The goal of the Network is to investigate mathematics from different theoretical and practical angles. The studies in this book fall into this interdisciplinary paradigm. We sincerely hope they provide insights into how mathematics is practiced and learned in the Information Age to both mathematicians and mathematics educators.

Toronto, ON, Canada   Stacy A. Costa
                                                                                                   Marcel Danesi
Windsor, ON, Canada   Dragana Martinovic

Contents

Chapter 1
The Information Age, Mathematics, and Mathematics Education

Stacy A. Costa, Marcel Danesi, and Dragana Martinovic

Introduction

Mathematical texts from across time indicate that the needs of the societies of different eras and different places have guided practices in mathematics itself and in how it was taught at school, responding to the needs and exigencies of each age. In some cases, even new discoveries were seen as part of a collaborative effort, rather than the product of individuals—the classic example being the Pythagoreans, who worked as a group to do mathematics (Heath 1921). A similar social attitude has emerged in the current Information Age, as evidenced by projects such as PolyMath, spearheaded by renowned mathematician Tim Gowers—a worldwide project involving mathematicians from all over the globe collaborating through the Internet to solve problems (Nielsen 2012). PolyMath started in 2009 when Gowers posted a famous problem on his blog, the density version of the Hales-Jewett theorem, asking people to help him find a proof for it. Seven weeks later, Gowers wrote that the problem had probably been solved, thanks to the many suggestions he had received. Since then, the PolyMath project has become a global collaborative project, recalling not only the ancient Pythagoreans but, in recent times, the Nicolas Bourbaki group of French mathematicians, who initially wanted to design updated textbooks

S. A. Costa
Department of Curriculum, Teaching & Learning / Collaboration in Engineering Education,
Ontario Institute for Studies in Education - University of Toronto, Toronto, ON, Canada
e-mail: stacy.costa@mail.utoronto.ca

M. Danesi
Department of Anthropology, Fields Institute for Research in Mathematical Sciences,
University of Toronto, Toronto, ON, Canada
e-mail: marcel.danesi@utoronto.ca

D. Martinovic (✉)
Faculty of Education, University of Windsor, Windsor, ON, Canada
e-mail: dragana@uwindsor.ca

© Springer Nature Switzerland AG 2020
S. A. Costa et al. (eds.), *Mathematics (Education) in the Information Age*,
Mathematics in Mind, https://doi.org/10.1007/978-3-030-59177-9_1

for teaching contemporary mathematics in the post-World War II era under this pseudonym, rather than under the name of any one individual.

Like the Bourbaki project, the findings, proofs, and discoveries of the PolyMath collaborators are published under a pseudonym, D. H. J. Polymath. Other projects similar to the PolyMath example have emerged, creating opportunities for new generations of mathematicians to work together and for students to develop their skills through collaborative ventures. At the same time, the discipline of computer science has come forth to provide new ways of doing mathematics, leading to a partnership between the two that is now a solid one.

These developments have brought about a paradigm shift in how mathematics is viewed, practiced, and taught. This chapter provides an overview of current Information Age mathematics, focusing specifically on: (1) how mathematics is evolving as a discipline on digital platforms (*crowdsourced mathematics*); (2) the implications of the partnership it has contracted with computer science (*computer mathematics*); and (3) what such trends imply for mathematics education today and possibly beyond. Our purpose is not just descriptive, but also analytical, since the break with previous traditions is a radical one and thus needs, at the very least, some cautious reflection. As the PolyMath project has shown, mathematics has found a powerful new socially-based way to do mathematics, called "massively collaborative mathematics," and this has had significant implications for mathematics education, which also require a critical assessment (Gowers and Nielsen 2009).

Crowdsourced Mathematics

On his blog, Gowers (2009) claims that progress in mathematics could be ensured more rapidly if mathematicians would work together. He puts it as follows:

> Why would anyone agree to share their ideas? Surely we work on problems in order to be able to publish solutions and get credit for them. And what if the big collaboration resulted in a very good idea? Isn't there a danger that somebody would manage to use the idea to solve the problem and rush to (individual) publication? Here is where the beauty of blogs, wikis, forums, etc. comes in: they are completely public, as is their entire history...If the problem eventually got solved, and published under some pseudonym like Polymath, say, with a footnote linking to the blog and explaining how the problem had been solved, then anybody could go to the blog and look at all the comments. And there they would find your idea and would know precisely what you had contributed. There might be arguments about which ideas had proved to be most important to the solution, but at least all the evidence would be there for everybody to look at. (Para. 8–9)

The PolyMath project envisions mathematics as social practice, rather than the efforts of individuals working alone. This view is ancient as the many anonymous ideas and proofs from antiquity attest. The difference today is that computers and the Internet have made it easy for collaboration to occur in a massive way. One example of this is through remote, and time shifting allowing for collaborative practices to occur anytime. The first such massive online collaboration is the Great

Internet Mersenne Prime Search (GIMPS), started by computer programmer George Woltman in 1996, whereby volunteers from across the globe search for larger and larger Mersenne primes, sharing information and insights. In addition to finding larger primes, which is a feat in itself, the project has provided a public forum for primality testing into which anyone can participate.

As Pease et al. (2020) have remarked, such online projects "provide us with a novel, rich, searchable, accessible and sizeable source of data for empirical investigations into mathematical practice," but also present various limitations associated with this approach, so-called *crowdsourced mathematics*, which is an example of what has come to be known as a social machine, defined as a combination of people and computers working on problems and projects as an entity. Similar to Star Trek's Borg Nation, consisting of cyborgs who are sharing a collective consciousness, as Shadbolt et al. (2019) have amply illustrated this has the potential to permanently change the way people do mathematics, and to transform the reach and impact of mathematics research.

As it is developing today in the Information Age, mathematics can be viewed from three main perspectives (Pease et al. 2020):

1. individual versus collaborative;
2. connected intelligence; and
3. human-computer interactions.

Mathematics has always been a deeply social discipline, as mentioned (Hersh 1998, 2014). The proofs of significant theorems have relied on the work of many mathematicians working alone, together, and in parallel. So, point (1) above is a moot one, since mathematicians have always collaborated, discussed, and debated proofs and solutions—the difference is that, in the past, the collaboration occurred slowly and in a scattered fashion via personal communication, usually in person and occasional group meetings and only in the last 200 years or so did it include mathematics journals and the responses they garnered. Nowadays, we witness a collaboration that is both massive and rapid. As Martin and Pease (2013: 1) point out, network collaborations through the Internet provide "a novel and rich source of data for empirical investigation of mathematical practice," and have provided us with "new ways to think about the roles of people and machines in creating new mathematical knowledge."

Point (2) encapsulates why the change in attitude towards mathematical practice has emerged. The notion of connected intelligence was elaborated in the mid-1990s by communications scholar Derrick de Kerckhove, referring to the effects of electronic environments on mindset (Kerckhove 1997). De Kerckhove argued that the Internet has allowed us to step outside the linearity of the previous print-literate brain. For de Kerckhove (1999), the electronically-connected world in which almost all of us live has provided a critical mass for the emergence of a connected form of intelligence, which means that the sum total of people's ideas is vastly more important than those of any one individual. He speculates that through this process we are undergoing one of the greatest evolutionary leaps in the history of our species. The architecture of connected intelligence resembles that of a huge brain whose cells and synapses are encoded in software and hardware that facilitate the assemblage of minds. Because of this, individual brains in the connectivity are able to see more, hear more, and feel more; but they also recede into the background. In this environment, experts are just members of the collective mind, whose ideas are carried by software and hardware systems that overlap with them and with relevant data and information. In effect, the emergence of connected intelligence is a result of living in an electronic age in which thoughts can travel through ether, and where "intelligence exists outside individuals and becomes pertinent once shared" (Kerckhove 1997: 4). As Marshall McLuhan (1998: 27) so aptly put it: "Man in the electronic age has no possible environment except the globe and no possible occupation except information-gathering."

Actually, this was anticipated both by Peter Russell in his 1983 book, *The global brain*, and even before that by philosopher Pierre Teilhard de Chardin in 1945. De Chardin's term for what is now "the Internet" and its connected intelligence structure, was the *noosphere*, a state of mind by which it would no longer be practicable to individuate the congeners of ideas or to assign importance to them. De Chardin saw this as part of the evolution of human consciousness, whereby individuals and collectivities are critical entities that are interconnected. The cliché "two heads are better than one" translates in this framework to "all heads are better than one."

It is important to note that De Chardin and McLuhan saw a danger in this collectivization force—namely, the loss of critical notions such as free will and the human spirit. So, they issued a warning that technological advances are human creations. They shape, not determine, how we communicate, interact, learn, and perceive ourselves; but they do not eliminate free will and the ability of the imagination to change things constantly. In other words, they judiciously warned that technology is our servant, not our leader. It was Spanish sociologist Manuel Castells (1996) who introduced the term *Information Age* broadly in his three-volume *The information age,* published between 1996 and 1998. Castells also argued that the digital technology and computer science revolutions have brought about unprecedented changes in the history of civilization, including a new form of collaboration that knows no traditional boundaries of geography, culture, or language, but at the same time has created conditions for exploitation of the individual and the loss of importance in individual choice.

Crowdsourced mathematics is a powerful new way to do mathematics, but it does not preclude the individual's participation in it as such, whether the person is named or not. Also, it cannot enhance mathematical creativity by default. One never knows when and to whom the spark will come. Consider a well-known anecdote that Henri Poincaré himself recounted in his book, *Science and method* (1908). Poincaré had been puzzling over an intractable mathematical problem, leaving it aside for a little while to embark on a geological expedition. As he was about to get onto the bus, the crucial idea came to him in a flash of insight. He claimed that without it, the solution would have remained buried somewhere, possibly forever. In effect, one cannot claim that one way (crowdsourcing) or the other (individual creativity) will bring about a discovery any better. As history shows, both are required.

Computer Mathematics

The second main area that has characterized mathematics in the Information Age is its partnership with computer science—point (3) above. Among the various accomplishments of this partnership, several stand out: (1) the creation of theorem-proving programs, (2) the use of computers by mathematicians to carry out proofs, (3) the use of computers to determine if a problem is provable and how quickly it can be solved, and (4) the use of computers to examine mathematical structures. Point (1) refers to automated theorem proving (ATP), which began in the mid-1950s with so-called first-order theorem provers, early computer programs designed to carry out proofs. One of the early first-order systems was able to prove 38 of the first 52 theorems of the Russell and Whitehead's, *Principia Mathematica* (Davis 2001). The idea of such programs was to emulate human proof-making. The objective of work on higher-order theorem provers was to devise programs that are not mimetic, but themselves innovative. Without going into details here, ATP aims to establish logical consequence at increasingly higher levels of proof using a computer. An ATP system is a powerful one if it can show that a theorem statement is (or is not) a logical consequence of the axioms or the propositional input. For an ATP system to be useful, it must also not be possible for it to prove non-logical consequences.

Point (2) refers to a heuristic use of computers to prove something. The first example of such a proof is the one for the four-color conjecture. As is well known, in its simplest form, this reads as follows: Is four the least possible number of colors needed to fill in any map, so that neighboring countries are always colored differently? The proof remained elusive for over a century when in 1976 at the University of Illinois at Urbana-Champaign, two mathematicians, Kenneth Appel and Wolfgang Haken (1977), put forward a proof that did not employ any of the traditional methods, but rather a computer program that could examine any map for the conjecture. The program found no map that required more than four tints to color distinctively and no exception to it has ever appeared. It has been called *proof by exhaustion*, because the computer algorithm devised for it has never produced an exception to the conjecture. The Haken-Appel proof constituted a true innovation in mathematical method.

Point (3) is a derivative of the previous two. As Fortnow (2013) argues, the problem of provability versus complexity, is a key one in computer mathematics. If one solves a 9-by-9 Sudoku puzzle, the task is a fairly simple one. The letter P is used to refer to this type of problem. The complexity arises when solving, say, a 1000-by-1000 puzzle. In this case, the symbol used is NP, which means that it would take more time for the computer to determine a solution and if it works. Computer algorithms can easily solve many complex Sudoku puzzles, but have difficulty as the degrees of complexity increase. The goal is to devise algorithms to find the shortest route to solving complex problems. If P were equal to NP then problems that are complex (involving large amounts of data) could be tackled easily as the algorithms become more efficient. The P = NP problem is the most important open problem in computer science, if not all of logic and mathematics. It seeks to know whether every problem whose solution can be quickly checked by computer can also be quickly solved by computer in polynomial time, where time is a simple polynomial function of the size of the input. It is becoming evident that even a sophisticated computer would take hundreds of years to solve some NP questions. Indeed, to prove P = NP one would have to use, ironically, one or more of the classic methods of proof.

An answer to the P = NP question would be to ascertain whether problems that can be verified in polynomial time can also be solved in polynomial time. Many problems can be checked quickly, but are slow to solve. One such example is finding prime factors of a large number. To check a solution, it is enough to multiply the prime factors, but to solve the problem—to find the prime factors of very big numbers—is very difficult, which is the basis for claiming that RSA (Rivest–Shamir–Adleman) encryption is considered to be secure.

Research has shown that a fast solution to a specific problem in NP can be used to build a quick solution to any other problem in NP—called NP-completeness. So far, it is not known if a fast solution will ever be found for NP-complete problems. This type of problem was mentioned by Kurt Gödel in a letter he sent to John von Neumann in 1956, asking him whether an NP-complete problem could be solved in quadratic or linear time (Fortnow 2013). The formal articulation of the problem came in a 1971 paper by Stephen Cook (and independently, a few years later by Leonid Levin 1973). Quadratic time refers to the fact that the running time of an algorithm increases quadratically if the size of the input is doubled. That is, as we scale the size of the input by a certain amount, we also scale the running time by the square of that amount. If we were to plot the running time against the size of the list, we would get a quadratic function.

Point (4) above (that is, the use of computers to examine mathematical structures) refers to writing a computer to model or simulate some solution, proof, or theory. It begins with a complete description of the operation that the computer is intended to model. This tells us what information must be inputted, what system of instructions and types of computing processes are involved, and what form the required output should take. The initial step is to prepare a model that represents the steps needed to complete the task. If the computer cannot handle a certain problem, the implication is that the problem would need to be studied further using other ideas and operations. In other words, if a model is inconsistent, the computer would be able to detect the inconsistency, because the program would go into a loop that lacks an exit routine. This is called *retroactive data analysis*—a method whereby efficient modifications are made to an algorithm and its correlative model that do not generate some output or at least do not correspond to the input data (see, for example, Demaine et al. 2004). The modifications can take the form of insertions in the model, deletions, or updates with new information and techniques. When nothing works, then the program has identified something that may be faulty in the model or, on the other hand, that may be unique to the phenomenon and thus non-computable, that is, beyond the possibilities of algorithmic modeling.

To summarize the foregoing discussion in a phrase, mathematics and computer science have formed a singular theoretical paradigm. The Information Age is a

Cybernetic Age. The term *cybernetics* was put forth in a 1948 book, *Cybernetics, or control and communication in the animal and machine,* by mathematician Norbert Wiener. For Wiener, certain mechanisms in machines serve the same purpose that aspects of the nervous system in humans serve, coordinating information to determine which actions will be performed. Their functions may differ, but their underlying *structure* is the same. It is relevant to note that Termini (2006) presents the history of cybernetics as one that fell short from developing into a full-fledged discipline. While cybernetics posed some excellent and quite novel questions that challenged the way people looked at humans and machines, its results were quickly appropriated by other disciplines. In that way, one super-discipline, that was meant to have a multidisciplinary core, was left without direction, losing ground to its disciplinary allies as they were gaining in visibility—Artificial Intelligence was adopted by Computer Science, Automata Theory and Formal Languages by Mathematics, and Control Theory by Engineering, as Termini (2006: 836) puts it:

> Cybernetics was unable to provide the deep unification it was aiming at; …This failure is so radical in so far as Cybernetics was meant not only to be a specific discipline (e.g. the science of control) but also a unifying paradigm. More precisely, it was seen as a new paradigm of scientific reason, as a unifying frame for other, already existing disciplines.

The Information Age can also be called the Computer Age, for obvious reasons—an age in which the computer has greatly amplified human capacities. It was Marshall McLuhan (1964) who suggested that all technologies are amplifications of human abilities. McLuhan framed this notion in the context of his *Four laws of media* (reported in McLuhan and McLuhan 1988)—amplification, obsolescence, reversal, and retrieval. A new technology or invention will at first amplify some sensory, intellectual, or other human psycho-biological faculty. While one area is amplified, another is lessened or rendered obsolete, until it is used to maximum capacity whence it reverses its characteristics and is retrieved in another medium.

A well-known, and now classic, example given by McLuhan (1962) is that of print technology. Initially, it amplified the concept of individualism because the spread of print materials encouraged private reading, and this led to the view that the subjective interpretations of texts was a basic right of all people, thus rendering group-based understanding obsolete until it changed from a single printed text to mass produced texts, leading to mutual readings, albeit typically displaced in time and space. This allowed for the retrieval of a quasi or secondary communal form of identity—that is, reading the same text connected readers in an imaginary way. This framework certainly seems to apply to the Computer Age, whereby the computer has greatly amplified the ability of mathematicians to do mathematics, while simultaneously making previously strict methods of proof obsolescent. Nonetheless, individual mathematicians continue to rely on classic methods as well to ply their profession, thus retrieving some of the historical precedents that have made mathematics the discipline that it is today.

Mathematics Education

The Information Age has had many implications for how mathematics education is delivered, given that the computer and the Internet have created conditions to amplify how mathematics is learned. Both crowdsourced and computer mathematics have now become virtually routine as methods for teaching mathematics. Because of these shifts, the "walls" of the past have started to crumble, with the mathematics classroom becoming increasingly a "wall-less" place, as McLuhan (1960) called it already in the late 1950s, since the computer can reach out beyond its traditional walled-in structure, both physical and academic. Mathematics can now be built upon and innovated with new methods all easily accessible and available to all.

Growing up in the Information Age, a typical student will see the partnership between mathematics and the computer as normal, rather than exceptional. McLuhan believed that the Industrial-Age model of mass education, with its walled-in classrooms, had passed, since the world was moving swiftly into a new era based on electronics and information, encouraging mutual involvement learning, while at the same time increasing individual creativity. McLuhan and Leonard (1967: 25) put it as follows:

> When computers are properly used, in fact, they are almost certain to increase individual diversity. A worldwide network of computers will make all of mankind's factual knowledge available to students everywhere in a matter of minutes or seconds. Then, the human brain will not have to serve as a repository of specific facts, and the uses of memory will shift in the new education, breaking the timeworn, rigid chains of memory may have greater priority than forging new links. New materials may be learned just as were the great myths of past cultures-as fully integrated systems that resonate on several levels and share the qualities of poetry and song.

It is mindboggling to consider that these words were written in 1967. The Internet has indeed made diversity of thought a concrete possibility, aiding us all in exploring and interacting with others in a global environment. We are all now expected to be "an explorer, a researcher, a huntsman who ranges through the new educational world of electric circuitry and heightened human interaction just as the tribal huntsman ranged the wilds" (McLuhan and Leonard 1967: 25).

With crowdsourcing possibilities, high school and college students now have an opportunity to collaborate creatively in ways that would have been unthinkable in the past—to interact with others outside their classroom to grasp certain mathematical concepts, to work with expert and authoritative sources, to collaborate on research projects and learning tasks with mentors and peer groups across the globe, etc. Also, with computer mathematics, the classic mathematical problems can now be easily studied through modeling, which is itself a powerful mode of deconstructing a problem and putting it back together. Consider, as a case-in-point, Alcuin's classic river-crossing problem, which he included in his medieval instructional manual, *Propositiones ad acuendos juvenes*, for training medieval youths in mathematics and logical thinking (Hadley and Singmaster 1992). There were actually four river-crossing problems in the manual, but the main one is the following, which is worth revisiting here for the sake of illustrating how it can be analyzed in terms of computer mathematics:

> A certain farmer needed to take a wolf, a goat and a head of cabbage across a river. However, he could only find a boat which would carry two of these [at a time], including himself. Thus, what rule did he employ to get all of them across unharmed, given that if he left the wolf alone with the goat, the wolf would eat the goat, and if he left the goat alone with the cabbage, the goat would eat the cabbage?

Alcuin had a pedagogical goal in mind when he created this problem. But little did he know the implications it harbored. The solution hinges on grasping the graphical structure involved in decision-making. The farmer cannot start with the cabbage, since the wolf would eat the goat if the two were left alone; nor the wolf, since the goat would eat the cabbage. So, his only choice is to start with the goat. Once this critical decision is made, the rest of the puzzle is solved easily. He goes across, drops off the goat, and comes back alone. When he gets back to the original side, he could pick up either the wolf or the cabbage. Let's go with the cabbage. He goes across with the cabbage to the other side, drops it off, but goes back to the original side with the goat (to avoid disaster). Back on the original side, he drops off the goat and goes over to the other side with the wolf. When there, he drops off the wolf to stay safely with the cabbage. He travels back alone to pick up the goat. He then travels to the other side with the goat and, together with the wolf and cabbage, continues on his journey. The graph presenting the solution to this problem is shown in Fig. 1.1:

The intellectual seeds of graph theory can be traced to this problem (Csorba et al. 2008). As Ito et al. (2012: 235) have shown, it has also implications for the P = NP problem, given that it is:

> NP-hard if the boat size is three, and a large class of sub-problems can be solved in polynomial time if the boat size is two. It's also conjectured that determining whether a river crossing problem has a solution without bounding the number of transportations, can be solved in polynomial time even when the size of the boat is large.

The problem thus lends itself in many ways to algorithm constructions, which bring out what it entails mathematically. In sum, this is an example of computer-based mathematics education. Given how much computers are now intrinsic to most people's lives, this approach also reflects discovery-based mathematics education, which allows students to grasp principles, such as graph-theoretic ones hidden within the river-crossing problem, by modeling them computationally.

McLuhan wrote in 1960 that the classroom of the future (from his times), would be a technologically-shaped one that is designed to open up the learning experience beyond the constrained environment of the traditional classroom, with tools that allow for self-sustained exploration on the part of the student. In the Information Age, this has certainly come about, as students and teachers alike use computers for various pedagogical purposes, and also communicate with others through cyberspace on a routine basis (Danesi 2016). Among the aspects that current mathematics education has amplified are the following:

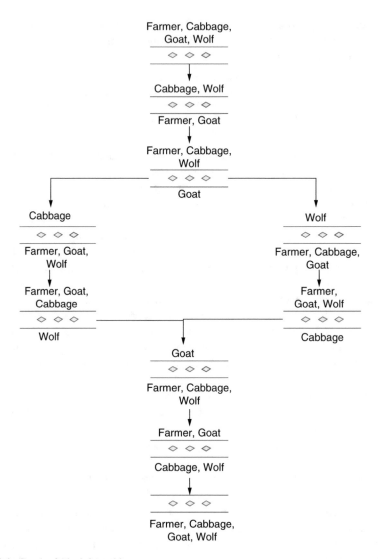

Fig. 1.1 Graph of Alcuin's problem

- Networking with other students and other teachers;
- Decentralization of methods and materials used;
- Increase in the speed and range of activities;
- Access to connected intelligence systems and crowdsourced mathematics; and
- Previous restrictions of time and space are eliminated.

Of course, even in previous ages, print materials allowed for access to, and sharing of, ideas beyond the classroom. But printed ideas move slowly since books and journals must be obtained, read, and then discussed, annotated, and studied in class or through further publication and perhaps via letter correspondence. So, while outside-the-classroom interaction was much slower, selective, and thus more remote, compared to current forms of interaction, it still allowed people to make contact constructively. The implication is clear—we must certainly keep moving more and more into the social structure of the Information Age, but we must do it judiciously, not with unbridled enthusiasm whereby anything from the past is considered essentially passé. Looking to the future, as McLuhan (1960) also pointed out, means looking to the past at the same time. While crowdsourcing, computer mathematics, and the like are exciting ways to learn, teach, and research mathematics, one should not forget history or break with it radically.

Concluding Remarks

The last comment leads us to Lewis Carroll. In his 1879 book, *Euclid and his modern rivals* (Carroll 1879), which takes the form of a comedic play, Carroll wanted to show the importance of Euclid in *a century when non-Euclidean geometries were taking a foothold in mathematics.* It is Carroll's defense of Euclid's *Elements* as the best textbook in geometry for a general audience, and an entry point into mathematics education itself. He encapsulates his objective in his introduction as follows (Carroll 1879: 1):

> The object of this little book is to furnish evidence, first, that it is essential, for the purpose of teaching or examining in elementary Geometry, to employ one textbook only; secondly, that there are strong a priori reasons for retaining, in all its main features, and specially in its sequence and numbering of Propositions and in its treatment of Parallels, the Manual of Euclid; and thirdly, that no sufficient reasons have yet been shown for abandoning it in favour of any one of the modern Manuals which have been offered as substitutes…In furtherance of the great cause which I have at heart—the vindication of Euclid's masterpiece—I am content to run some risk; thinking it far better that the purchaser of this little book should read it, though it be with a smile, than that, with the deepest conviction of its seriousness of purpose, he should leave it unopened on the shelf.

Carroll's advice is still a valid one today, albeit more as a warning than anything else. The question is not whether or not to insert Euclid's *Elements* in the curriculum as a means for grasping mathematics; but rather, for being wary of new trends in themselves. As mathematics education continues to change in the Information Age, it must also keep an eye on the past and not amputate it—a term McLuhan used to great effect. Studying Euclid today would not be anachronistic if the *Elements* are inserted into an eclectic mathematics curriculum. There is a real danger that the modern world may be amputating the past more and more. Bringing the past into the present and future is easily accomplished, as the river-crossing problem above was meant to illustrate. Together with current technologies, its use in the classroom allows for such amalgamation of the past and present:

1. *Modeling*: The puzzle provides a means to experiment with its structure through computer modeling.

2. *Generalization*: It opens up opportunities for connection with others online so as to go beyond the puzzle itself and seek ways to generalize its structure.
3. *Extension*: Connecting with others online will allow learners to explore other implications of the puzzle.
4. *Amalgamation:* This involves amalgamating various solutions, perspectives, models, etc. of the puzzle into a mini-treatise of its implications.
5. *Multiplicity:* This entails discriminating and grasping multiple perspectives on the puzzle and its mathematical structure.
6. *Transmedia navigation:* This entails knowing how to follow the flow of ideas, events, and information across multiple media sites based on the puzzle.
7. *Judgment:* This means developing the ability to discern what is legitimate or not.
8. *Collective Intelligence:* This involves understanding how to pool knowledge and collaborate with others towards common objectives.

While all of these aspects of learning existed somewhat in previous epochs, the revolution in connective technology has brought about a new sense that understanding and communicating can no longer be constrained to the printed page or the traditional walled-in classroom. And this has had enormous repercussions, as McLuhan (1996: 275) predicted decades ago:

> We now live in a technologically prepared environment that blankets the earth itself. The humanly contrived environment of electric information and power has begun to take precedence over the old environment of nature. Nature, as it were, begins to be the content of our technology.

If we keep this in mind, it is obvious that the Information Age has led to a new stage in the study of mathematics and its teaching. It has indeed made diversity of thought a concrete possibility, aiding us all in exploring and interacting with others in a global environment. We are all now expected to be "an explorer, a researcher, a huntsman who ranges through the new educational world of electric circuitry and heightened human interaction" as McLuhan and Leonard (1967: 25) pointed out, rather prophetically, decades ago. Only time will tell if the Internet of today has brought us closer to *noosphere*, which de Chardin envisioned as driven by generosity and friendship, or if it has brought about the loss of free will and human spirit as in the Borg nation. Despite the encouraging examples that we have mentioned here, we should not forget that people can get lost in a crowd, and individual voices may not be heard among a multitude of shouters, and that our youth may lose direction when confronted with too many choices (Noveck 2000). It may be up to educators to realize the full potential of increasingly networked societies and to stand at the forefront of a movement that will shape our future (Martinovic and Magliaro 2007).

References

Appel, K. and Haken, W. (1977). Every planar map is four colorable. *Illinois Journal of Mathematics* 21: 429–490.
Carroll, L. (1879). *Euclid and his modern rivals*. London: Macmillan.

Castells, M. (1996). *The information age: Economy, society, and culture.* Oxford: Blackwell.

Cook, S. (1971). The complexity of theorem proving procedures. *Proceedings of the Third Annual ACM Symposium on Theory of Computing,* 151–158. doi:https://doi.org/10.1145/800157.805047.

Chardin, T. de (1945). *The phenomenon of man.* New York: Harper.

Csorba, P., Hurkens, C. A. J., and Woeginger, G. J. (2008). The Alcuin number of a graph. *Algorithms: ESA 2008, Lecture Notes in Computer Science, 5193*: 320–331.

Danesi, M. (2016). *Learning and teaching mathematics in the global village: Math education in the digital age.* New York: Springer.

Davis, M. (2001). The early history of automated deduction. In: A. Robinson and A. Voronkov (eds.), *Handbook of automated reasoning.* Oxford: Elsevier.

Demaine, E. D., Iacono, J., and Langerman, S. (2004). Retroactive data structures. *Proceedings of the 15th Annual ACM-SIAM Symposium on Discrete Algorithms, 2004*, 274–283.

Fortnow, L. (2013). *The golden ticket: P, NP, and the search for the impossible.* Princeton: Princeton University Press.

Kerckhove, D. de (1997). *Connected intelligence: The arrival of the web society.* Toronto: Somerville.

Kerckhove, D. de (1999). Connected Intelligence. The Information Management Round Table. Retrieved from https://www.researchgate.net/publication/30871546_Connected_Intelligence.

Gowers, T. (2009). Is massively collaborative mathematics possible? *Gowers's Weblog.* Retrieved from https://gowers.wordpress.com/2009/01/27/is-massively-collaborative-mathematics-possible/.

Gowers, T. and Nielsen, M. (2009). Massively collaborative mathematics. *Nature* 461: 879–881.

Hadley, J. and Singmaster, D. (1992). Problems to sharpen the young. *Mathematics Gazette* 76: 102–126.

Heath, T. L. (1921). *A history of Greek mathematics.* Oxford: Oxford University Press.

Hersh, R. (1998). *What is mathematics, really?* Oxford: Oxford University Press.

Hersh, R. (2014). *Experiencing mathematics.* Washington, DC: American Mathematical Society.

Ito, H., Langerman, S. and Yoshida, Y. (2012). Algorithms and complexity of generalized river crossing problems. In: E. Kranakis, D. Krizanc, and F. Luccio (eds.), *Fun with algorithms. Lecture notes in computer science,* 235–244. Berlin: Springer.

Levin, L. A. (1973). Universal search problems. *Problems of Information Transmission* 9: 115–116.

Martin, U. and Pease, A. (2013). Mathematical practice, crowdsourcing, and social machines. *Intelligent computer mathematics. MKM, Calculemus, DML, and systems and projects 2013, held as part of CICM 2013,* Bath, UK, July 8–12, 2013.

Martinovic, D., and Magliaro, J. (2007). Computer networks and globalization. *Brock Education Journal* 16: 29–37.

McLuhan, M. (1960). Report to the National Educational Broadcasters Association: In: P. Marchand (ed.), *Marshall McLuhan: The medium and the messenger.* Cambridge, Mass.: MIT Press.

McLuhan, M. (1962). *The Gutenberg galaxy: The making of typographic man.* Toronto: University of Toronto Press.

McLuhan, M. (1964). *Understanding media: The extensions of man.* Cambridge, Mass.: MIT Press.

McLuhan, M. (1996). Culture without literacy. In: E. McLuhan and F. Zingrone (eds.), *Essential McLuhan.* New York: Basic Books.

McLuhan, M. (1998). *The agenbite of outwit,* published posthumously in *McLuhan Studies,* Volume 1, Issue 2 (January 1998).

McLuhan, M. and Leonard, G. B. (1967). The future of education: The class of 1989. *Look,* February 21, pp. 23–24.

McLuhan, M. and McLuhan, E. (1988). *Laws of media: The new science.* Toronto: University of Toronto Press.

Nielsen, M. (2012). *Reinventing discovery: The new era of networked science.* Princeton: Princeton University Press.

Noveck, B. (2000). Paradoxical partners: Electronic communication & electronic democracy. In: P. Ferdinand (ed.), *The Internet democracy and democratization.* 18–36. Ilford, Essex: Frank Cass.

Pease, A., Martin, U., Tanswell, F. S., and Aberdein, A. (2020). Using crowdsourced mathematics to understand mathematical practice. *ZDM Mathematics Education.* doi:https://doi.org/10.1007/s11858-020-01181-7.

Poincaré, H. (1908). *Science and method.* New York: Dover.

Russell, P. (1983). *The global brain.* New York: Tarcher.

Shadbolt, N., O'Hara, K., De Roure, D., and Hall, W. (2019). *The theory and practice of social machines.* New York: Springer.

Termini, S. (2006). Remarks on the development of cybernetics. *Scientiae Mathematicae Japonicae Online*, e-2006, 835–842.

Wiener, N. (1948). *Cybernetics, or control and communication in the animal and the machine.* Cambridge, Mass.: MIT Press.

Chapter 2
An Unsupervised Approach to User Characterization in Online Learning and Social Platforms

Dan Vilenchik

A Short History of User Characterization

Making sense of data that is automatically collected from online platforms such as online social media or e-learning platforms is a challenging task: the data is massive, multidimensional, noisy, and heterogeneous (composed of differently behaving individuals). In this chapter we focus on a central task common to all on-line social platforms and that is the task of user characterization. For example, automatically identify a spammer or a bot on Twitter, or a disengaged student in an e-learning platform.

Understanding the nature and patterns of interaction between members of a social network is a long standing research topic. Back in the 1950s (Katz and Felix Lazarsfeld 1957) studied the problem of identifying influential people in social networks. Two decades later, Freeman's seminal work (Freeman 1978) coined three key indices of centrality: degree (the number of friends), closeness (the average number of hops from a user to all other users in the network) and betweenness (the fraction of shortest paths that have to go through this user), fueling a torrent of theoretical and experimental work in the area. The subject became even more attractive to researchers and industry as the role of *online* social networks (OSNs) increased dramatically in recent years, with new business opportunities for marketeers.

The task of characterizing users of OSNs is typically approached as a supervised learning classification problem. A target variable is defined, e.g. the ethnicity and political affiliation of a user (Pennacchiotti and Popescu 2011), gender, age, regional origin (Rao et al. 2010), occupational class (Preotiuc-Pietro et al. 2015), etc. Next, data is collected from the network (typically using some sort of crawling procedure), and relevant features are extracted from each user account. Finally, one of a host of machine learning algorithms is trained to perform the task.

D. Vilenchik (✉)
School of Electrical and Computer Engineering, Ben-Gurion University, Beersheba, Israel
e-mail: vilenchi@bgu.ac.il

© Springer Nature Switzerland AG 2020
S. A. Costa et al. (eds.), *Mathematics (Education) in the Information Age*,
Mathematics in Mind, https://doi.org/10.1007/978-3-030-59177-9_2

The task of characterizing types of students in e-learning systems shares many similarities with that of users of OSNs. One central type of characterization is the level of engagement of a student with the system. This is a challenging task as the nature of students' engagement in self-directed e-learning systems is in general quite variable. Students may return to the system after prolonged periods of absence and vary widely in the types of activities they choose, and in their level of competence. Some works partition users to engagement types according to predefined notions such as course completion rate or time spent in the system (Cocea and Weibelzahl 2007; Lloyd et al. 2007). Others have viewed student engagement as a multidimensional construct, that involves different factors such as student generated content, social interaction and their learning outcomes (Ramesh et al. 2013, 2014). Significant attention in the literature has been attributed to modeling disengagement in online education systems by tracking how student performance changes over time and by predicting dropouts (Crossley et al. 2016; Lykourentzou et al. 2009; Yang et al. 2013).

One drawback of using a supervised learning approach is that obtaining labeled data for training a classifier may be highly non-trivial or costly, both in e-learning systems and in online social platforms. Furthermore, when deciding the target variable upfront, a relatively narrow view of the platform and its users is obtained. For example, if one is to classify engagement, then one has to commit upfront to a certain definition of engagement (time in the system? performance? throughput?) For those reasons, the capacity of unsupervised learning algorithms for user characterization was studied as well. There are two main methods of finding patterns in data in an unsupervised manner: clustering and low-dimensional approximation methods such as Principal Component Analysis (PCA). The latter, will be our focal point. In this chapter we discuss in detail how PCA may be used to characterize users of both OSNs and e-learning platforms. We cover both the theoretical tenets of the methodology and two in-depth real-data examples. Finally, we discuss a surprising Simpson-like paradox which in some cases is coupled with the PCA-based method.

Methodology: Using PCA to Characterize Users

For completeness we start with a brief overview of PCA and how it is used to characterize users.

Notation

We shall use bold lower-case letters, e.g. $\mathbf{x}$, to signify vectors, and non-bold lower-case letters, e.g. x, to signify scalars. Upper-case letters are reserved for matrices. We consider all vectors as column vectors. We let p be the number of features collected for each user, n the number of users in the sample and X the resulting

$n \times p$ data matrix. We let $\hat{\Sigma} = \dfrac{1}{n} X^T X$ be a sample covariance matrix of the design matrix X.

A Crash Course on PCA

Principal components analysis (PCA) is the mainstay of modern machine learning and statistical inference, with a wide range of applications involving multivariate data, in both science and engineering. PCA is mostly used as a tool in exploratory data analysis, allowing visualization of the data by projecting it into a carefully chosen lower-dimensional space.

Recalling the derivation of PCA, which can be found for example in (Anderson 1962), the first PC is the direction (unit vector) $v_1 \in R^p$ in which the variance of X is maximal. Standard algebraic manipulations show that v_1 is also the leading eigenvector of 's covariance matrix, $\hat{\Sigma}$. This gives a convenient algorithmic way to compute v_1. The remaining PCs, $v_2, \ldots, v_p$ are defined in a similar way and together they form an orthonormal basis of R^p. Some additional algebraic manipulations give that the percentage of variance explained by v_i is simply $\lambda_i / \left(\sum_{j=1}^{p} \lambda_j \right)$, where λ_i is the eigenvalue associated with v_i.

Under various reasonable assumptions, when n, the number of samples, is much larger than p, the number of features, then the PCs indeed point at the "true" direction of variance (true in the sense that it fits the latent underlying distribution according to which the data is distributed) (Anderson 1962; Muirheads 2005). This is indeed the typical case both in OSNs and e-learning platforms.

Characterizing Users with PCA

Suppose, for simplicity of presentation, that every data point is represented using two features f_1, f_2. In other words, every data point $\mathbf{x}$ may be expressed as $\mathbf{x} = (f1, f2) = f1 \cdot (1,0) + f2 \cdot (0,1)$. Thus, the features f_1, f_2 correspond to the standard axes $(1,0)$ and $(0,1)$ of R^2. Every PC v_i is in particular a vector $v_i = \left(v_{i_1}, v_{i_2} \right) = v_{i_1} \cdot (1,0) + v_{i_2} \cdot (0,1)$. Thus, the PCs are linear combinations of the original set of features. As such they may be interpreted as a new set of complex features, forming a new set of axes along which the data is redrawn. The value of each data point $\mathbf{x}$ in the new coordinate system is given by the scalar product $\langle v_i, \mathbf{x} \rangle$ along the i th axis. In this manner the PCs induce a soft classification of the users. For example, if the leading PC v_1 is a linear combination of features that indicate that the user is popular (say, the number of retweets of that user's posts, the number of likes the user receives, etc.), then the larger a user's projection on v_1 the more popular the user is.

For a PCA-based characterization scheme to work, the directions at which the PCs point, need to be amicable for semantic interpretation. How such an interpretation process is typically carried out? Ideally, one first looks at the features f_i where the weight in the linear combination is non-zero (we will call this set the support of v and denote it by $supp(v)$). Take all the features that were selected and see if a natural label/quality can be assigned to them, e.g. "spammer", "bot", "highly-engaged student", etc. Unfortunately, every PC v will typically satisfy $supp(v) = \{1, \ldots, p\}$, yet because numerical reasons, making it impossible to interpret any of the PCs. One popular solution is the widely-used "interpret-by-top-k" rule. The rule says first to sort the PC vector entries in descending order of absolute values, then assign the PC its label according to the top k features, ignoring entries with smaller values. While this practice is useful in many cases, the choice of k is subjective and may affect the interpretation. In addition, choosing smaller k values makes interpretation easier, as fewer features are involved, but possibly at the cost of semantic validity. We refer the interested reader to (Vilenchik et al. 2019) for a detailed discussion.

To interpret the PCs, We follow the framework suggested in (Vilenchik 2020), which is a "safe" variant of the "interpret-by-top-k", circumventing the caveat of choosing the right k. For simplicity of presentation assume that the features that were collected fall into two categories, not necessarily disjoint, that may be captured by two qualities Q_1, Q_2 (e.g. Q_1 is the quality of being popular and Q_2 the property of being a spammer). For every PC v_i we define its *energy* in the direction of Q_1 and Q_2 as follows:

$$\alpha_i = \sum_{r \in Q_1} \left(v_i[r] \right)^2, \quad \beta_i = \sum_{r \in Q_2} \left(v_i[r] \right)^2. \tag{2.1}$$

The total energy of every v_i is 1 as it is a unit vector. Hence $\alpha_i, \beta_i \in [0, 1]$. The ideal scenario with respect to interpretability is equivalent to requiring that for every PC v_i, either $\alpha_i = 1$ or $\beta_i = 1$. We may replace this ideal requirement by the relaxed requirement that $\alpha_i \geq x^{\alpha}_{0.975}$ and $\beta_i \leq x^{\beta}_{0.025}$ or vice versa, where $x^{\alpha}_{0.975}$ is the 97.5%-percentile of the α value had the vector v_i been a random p-dimensional vector on the unit sphere. We call this the (α, β)-*separation property*. The choice of 0.975 and 0.025 is somewhat arbitrary, and in general the closer the percentiles are to 100% or 0% the closer we are to the ideal setting. Finally, every PC that satisfies this property, may safely be classified as Q_1 (if $\alpha > \beta$) or Q_2 (if $\beta > \alpha$).

Typically, only PCs that explain a significant percentage of the variance are considered for interpretation, while the rest are being ignored. This is commonly known as the Kaiser-Guttman criterion (Yeomans and Golder 1982). We discuss it in more detail in the next sections where we present concrete examples.

Using PCA in a Dynamic Fashion

So far we have seen how PCA is used in a static way, namely, the PCs are computed, and users are characterized by the extent of their projection along each PC. However, the axes provided by PCA can be used to generate a dynamic time-dependent view from which yet more patterns may be obtained. We follow (Hershcovits et al. 2020) in the description of this dynamic framework.

A time series for user u consists of projections of its accumulated data on a fixed PC at a fixed frequency. Formally, with every user u at time t, a vector $\boldsymbol{u}_t$ is associated, which consists of the current value of the p features that are measured for each user. For every user u and PC $\boldsymbol{v}_i$, a time series $S_u^{(i)} = \left\{ \alpha_{t_1}^{(i)}, \alpha_{t_2}^{(i)}, \ldots \right\}$ is produced, where $\alpha_{t_j}^{(i)}$ is the projection of $\boldsymbol{u}_{t_j}$ in the direction of $\boldsymbol{v}_i$. The time-series are then labeled according to the trend of their graph. We will soon see how this labelling is anchored in a theory of education:

- *Fixed* if all $\alpha_{t_j}^{(i)}$ are the same
- *Monotone Up* if $\alpha_{t_j}^{(i)} < \alpha_{t_{j+1}}^{(i)}$ for all time stamps. *Monotone Down* is defined symmetrically.
- *Variable* if it is neither fixed nor monotone.

The users are then clustered according to some assignment rule which is a function of the labels assigned to their time series. For example, all users whose time-series on the leading PC $\boldsymbol{v}_1$ is Fixed or Monotone belong to cluster A. All users with a variable time-series for $\boldsymbol{v}_1$ are in cluster B. Finally, one would assign a meaningful quality to each cluster, thus obtaining user characterization.

Let's continue our example and take for concreteness some e-learning domain, and the quality will be the level of engagement of a student as measured by the total time spent in the system. The relevant working hypothesis is that diverse use of the system leads to prolonged student engagement. This notion is supported by studies in e-learning showing that varying the pedagogical challenges posed to students, and skills requires to meet the challenges, positively contributes to students' motivation and retention (Pearce 2005; Rodrıguez-Ardura and Meseguer-Artola 2017). Using the variability of a student's time-series trajectory as a proxy for diverse engagement, cluster A of the Fixed and Monotone series will be labeled as the less engaged students, and cluster B of Variable, will be labeled as the highly engaged. In the next section we repeat this example with the complete detail.

Example 1: Using PCA to Characterize Students in an e-Learning Platform

Let us now turn to see how the methodology that we discussed so far is implemented on real data. Our first example is based on (Hershcovits et al. 2020), where the interested reader may find the full details.

The data, 18,979 users who were active in the system between June 2015 and January 2016, was collected from an online math learning platform for K-9 education. The platform is available on the web and hand-held devices and used throughout the world. It is based on a set of interactive games, called episodes. Each episode is a game for practicing a mathematical skill (e.g. counting) and consists of 5–6 questions. The platform contains over a thousand episodes, each tagged with a math skill and suitable grade level, designed to convey mathematical concepts to students in a way that promotes self-discovery and skill acquisition. Students choose episodes at will, or have episodes from a particular skill set or grade level assigned to them at random.

Despite the success in registering new users to the system, it exhibits a high attrition rate. More than 11% of users did not complete a single question, 37% of users completed ten or less questions; 67% of the users spent less than 100 min in the system.

The raw data from the system consists of logs for the following events: a user opening and closing an episode, submitting a response for a question in an episode, time taken to submit response, whether the response was correct. 17 features were designed based on these events; Table 2.1 below provides the feature map.

Table 2.1 Description of features collected for every user in the e-learning platform (Hershcovits et al. 2020)

Feature name	Description
EpisodeMaxDiff	Most difficult episode that the user opened
EpisodeAvgDiff	Average episode difficulty for that user
EpisodeStdDiff	Standard deviation of EpisodeAvgDiff
QuestionFalseRate	Percentage of incorrect question responses
NumQuestions	Total number of questions solved by the user.
CompletionRate	Percentage of episodes that were completed
NumEpisodes	Total number of episodes opened by the user
RepetitionRate	The percentage of episodes opened twice consecutively by the user
MostRecentCorrect	Boolean value, true if most recent answer was correct
EpisodeMinDiff	The easiest episode that the user opened
AvgResponseTime	Average time between responses to consecutive questions in an episode
STDResponseTime	Standard deviation of AvgResponseTime
STDLagTimeEpisodes	Standard deviation of AvgLagTimeEpisodes
LastReponseTime	Response time to the last question in the last episode
AvgLagTimeEpisodes	Average time between completion of episode and beginning of next.
DistinctEpisodesRate	Fraction of distinct episodes that were opened
AbandonRate	The percentage of episodes that were opened but not completed

PCA: A Static View

First, PCA was computed for the 17×17 covariance matrix of the roughly 19,000 users. Only the top five PCs explain more than $1/p = 1/17$- fraction (around 5%) of the variance, the rest are ignored the Kaiser-Guttman criterion (Yeomans and Golder 1982). Table 2.2 gives the top three PCs weights (aka loadings) with respect to every feature.

Let us now demonstrate how the top three PC's may be interpreted. We defer the exact computation of energy, using Eq. (2.1), to the next example, and instead provide here a more intuitive description.

The features providing the largest contribution to the support of the leading PC v_1 are the maximum and average difficulty of the episodes opened by the user, the standard deviation of the difficulty of the episodes (both in positive sign), the abandon rate and the distinct episode rate (in negative sign). From this we may conclude that the quality that v_1 points to are students that solve episodes with a variety of difficulty levels and make multiple attempts to solve the episodes.

For concreteness, we examine a user from the database whose projection on v_1 was in the 90%-percentile. This user opened 19 episodes, but only 11 of these episodes were distinct episodes (58% distinct rate, compared to 88% on average for all users). This user experienced with different level of episodes' difficulties (standard deviation difficulty rate of 20% vs. 9% SD averaged over all the users), with max difficulty of 79%, min of 18% and average of 37%. The user's abandon rate was 21% (compare to 31% in average).

Table 2.2 Loadings for top three PCs v_1, v_2 and v_3 in the e-learning system (Hershcovits et al. 2020)

Feature	v_1	v_2	v_3
EpisodeMaxDiff	0.42	0.04	−0.11
EpisodeAvgDiff	0.36	0.35	−0.18
EpisodeStdDiff	0.33	−0.08	−0.042
QuestionFalseRate	0.27	0.24	−0.17
NumQuestions	0.25	−0.48	−0.01
CompletionRate	0.25	−0.04	0.10
NumEpisodes	0.24	−0.48	−0.01
RepetitionRate	0.21	−0.07	−0.08
MostRecentCorrect	0.18	0.03	0.12
EpisodeMinDiff	0.13	0.50	−0.17
AvgResponseTime	0.12	0.14	0.58
STDResponseTime	0.11	0.06	0.56
STDLagTimeEpisodes	0.11	−0.09	−0.013
LastReponseTime	0.08	0.09	0.45
AvgLagTimeEpisodes	0.06	−0.01	−0.02
DistinctEpisodesRate	−0.27	0.19	0.078
AbandonRate	−0.34	−0.13	−0.07

Moving on to v_2, the top largest entries correspond to the episode difficulty rate (minimum and average), the false rate (in positive sign), the number of questions and the number of episodes played (in negative sign). Therefore, the quality that v_2 points to are students that attempt to solve hard episodes, have a high false rate in these episodes, and a low throughput in the system. Indeed, the quality suggested by v_2 is very different than that of v_1. To illustrate, we look at a user with whose projection on v_2 was in the 99%-percentile. This user answered only 8 questions (compared to 61 on average for all users), in 4 episodes opened (compare to 13 on average). The user chose very difficult episodes (in the 90%-percentile), and had false rate of nearly 100%.

Finally, looking at v_3, the largest entries are only positive and correspond to the aspect of time: the average response time, the standard deviation of the response time and the last response time. Hence v_3 points at the quality of a student that spends more time answering questions, but with large variation in response times.

Now we take a user with a small projection on v_3, below the 10%-percentile. The user answered 12 question, with average response time (normalized) of 0.25 (compared to 0.91 on average), and standard deviation of 0.08 (compare to 0.97 on average). In addition, the user's false rate was 17% which is quite similar to the average of 22%, therefore it might be that the user was mainly guessing the answers.

PCA: A Dynamic View

The next type of patterns which may be drawn from the PCs come from the geometric trends of the time-series. In this specific example, for every user three time series were generated, one for each of the top three PCs. The sample rate was every 10 min.

Table 2.3 summarizes the percentage of users that belong to each of the four trajectory labels, the total time in the system and the net activity time for each label. Indeed, the table corroborates the working hypothesis: diverse use of the system leads to prolonged user engagement. To see how this working hypothesis manifests concretely, let us portray various possibilities for time-series along v_1. The PC v_1 is mainly supported by features that correspond to question difficulty and variability in difficulty. The Fixed label captures users that did not solve any episodes (and the projection is fixed to 0) or solved one episode before quitting. Their time in the system is the lowest, and their activity is least diverse. There are several possibilities for a Monotone Down trajectory: either the user starts with difficult questions and gradually lowers the difficulty. Another option is that the abandon rate increases. We can summarize this trajectory as a user that fails to calibrate on a suitable level of difficulty and drops from the system. Monotone Up users experience exactly the opposite: they solve harder and harder episodes, and experience with various levels of difficulty. We see that their average time in the system is almost 7 times higher than Monotone Down. Users in the Variable trajectory experience with various levels of difficulty, which may suggest a successful calibration of difficulty level. As a result, they stay the longest in the system (4 times more than Monotone Up).

Table 2.3 Breakout of users according to trajectory type of time-series generated for v_1 (top), v_2 (middle) and v_3 (bottom)

v_1	Percentage	Time in system (hours)	Net activity time (min)
Fixed	36%	0.11	1.8
Monotone up	17.3%	81.3	17.6
Monotone down	5.3%	12.1	11.2
Variable	41.4%	363.5	107.8
v_2	Percentage	Time in system (hours)	Activity time (min)
Fixed	36%	0.12	1.8
Monotone up	6%	15.2	14.2
Monotone down	17.7%	95.3	19.1
Variable	40.3%	365.6	109.2
v_3	Percentage	Time in system (hours)	Activity time (min)
Fixed	36%	0.12	1.8
Monotone up	23%	23.9	15.1
Monotone down	10.4%	74.6	11.9
Variable	44.5%	348.4	102.5

In addition, the total time and the net activity time are presented (Hershcovits et al. 2020)

Clustering Users

The different shapes of the time series may be used to assign users into clusters (or cohorts). The assignment rule depends on some domain expert knowledge. In our example, it's the working hypothesis that diversity of use leads to prolonged engagement.

Here is a possible assignment rule that is consistent with the hypothesis: if the time series is Fixed or Monotone assign A, and if Variable assign B. This rule differentiates between what's known in the literature as *Early droppers* vs. *Intermediate/ Fully adopters*. Furthermore, other, more sophisticated, rules may be used for clustering, for example, assign B only if the variability level (say, measured as the number of times the time series graph changes trend) exceeds a certain threshold.

A natural question that arises is how to validate the usefulness of a clustering rule? One possible validation test is the following. Collect data on all users up to a certain time T. Apply the clustering rule only according to the data collected up to time T. Then, check what percentage of users that were assigned to cluster A are indeed early droppers, namely they dropped out of the system before or shortly after time T, and what percentage of cluster B users are fully adopters, namely the stayed in the system considerably longer than time T.

The results of this validation experiment, once with $T_1 = 10$ min (sampling at a 1-min frequency) and second with $T_2 = 5$ days (sampling at 10 min frequency) are summarized in Table 2.4. Three versions of the aforementioned assignment rule were examined (PCA_v1, PCA_v2, PCA_v3), each rule with a different variability level threshold. In addition, standard off-the-shelf supervised-learning classifiers were trained to obtain a baseline for comparison.

Table 2.4 AUC for the two prediction tasks using tenfold cross validation

	RF	LDA	NB	PCA_v1	PCA_v2	PCA_v3
Task 1	81.1%	81.5%	77.5%	73.9%	80.5%	81.2%
Task 2	68.8%	72.2%	69.3%	65.5%	69.6%	69.1%

Task 1: Cutoff at T = 10 min, Task 2: cut-off at T = 5 days. The best two classifiers are highlighted (Hershcovits et al. 2020)

As customary in unbalanced classification problems (i.e. the two classes differ significantly in size), the AUC is taken as the gold standard criteria. Table 2.4 presents the AUC value of the PCA-based classifier alongside the supervised-learning classifiers. The performance of the PCA-based classifiers is similar (if not better) than the supervised learning algorithms. However, there are two major benefits in using the time-series PCA-based prediction. The first, obviously, is the fact that no labeled data is required to train the classifier. The second is the fact that the decision about engagement type is based on domain-specific theories (like the working hypothesis that we presented) rather than a decision made by a black-box classifier.

Example 2: Using PCA to Characterize Users in OSNs

Our next example draws on various results that were obtained in recent years where PCA was used to characterize users in a variety of online social platforms, in a very similar manner to what we have just presented for the e-learning platform. So far in OSNs, PCA was only used in its static form and not as a basis for a time-series clustering.

We start by explaining how typically OSN-data is obtained. The network is crawled in a snowball approach, which is commonly used in the literature (Mislove et al. 2007). Crawling starts from a list of randomly selected users and proceeds in a BFS manner. At each step the crawler pops a user v from the queue, explores its outgoing links and adds them to the queue. In Twitter for example, there is a link from v to w if v follows w. In other networks, Facebook for example, the friendship relationship is symmetric.

Our example builds on the data that was collected in two papers (Canali et al. 2012; Vilenchik et al. 2019). Six OSNs were crawled: YouTube, Flickr (Canali et al. 2012), Twitter, Instagram, Flickr and Steam (Vilenchik et al. 2019). For each platform, between 9 and 15 features were collected. The features vary from platform to platform, but in general they include two categories of features. Feedback features, which included for example the number of users following me, the number of retweets of my tweets by others, the number of likes I received or comments left on my pictures.

The second category is Activity features, which included the volume of activity (e.g. posts per day, total number of posts), activity types (e.g. percentage of video vs pictures, urls vs. pure text), social activity (number of friends, number of likes I gave, number of tweets I retweeted). Similar features were collected for example in (Eirinaki et al. 2012) to find influential users in MySpace and Facebook. The complete set of features can be found in the two aforementioned papers. We included Tables 2.5 and 2.6 from (Vilenchik et al. 2019) for completeness, and they detail the features of Twitter and Steam respectively.

We selected to explore in depth two platforms, Twitter and Steam (an on-line gaming platform), as they exemplify to antipodal scenarios of a PCA-based characterization scheme: In Twitter the PCs could be usefully interpreted while in Steam no interpretation could be assigned to any of the top PCs.

Table 2.7 details the percentage of variance that each PC explains, and the PCs that explain more than $1/p$-fraction of the variance (the Guttman-Kaiser criterion (Yeomans and Golder 1982)) can be observed. Next, similarly to the e-learning example, we look at the PC loadings table, Table 2.8 for Twitter and Table 2.9 for Steam, and see if the energy of the PCs is centered on a specific quality: Feedback, which we may identify with popularity, or Activity.

Tables 2.10 and 2.11 summarize the percentage of energy that each quality possesses in each of the six OSNs. We see that some PCs are pure and some are not. In the next section we discuss what does it mean that in some platforms, all PCs are pure, and in some none. But for now, let us go back to Twitter and interpret the leading top three PCs according to the loadings table and the energy table.

The leading PC v_1 is pure Feedback, supported by the three features *NumOfFollowers, NumOfOtherRT* and *LikesGivenToMe*, which all count feedback. To corroborate that $\mathbf{v}_1$ is a popularity measure one may also look at the crawl sample, and consider the users with the largest projection on that PC. Indeed, one finds A-list celebrities. As the projection diminishes, the A-list celebrities make room to local ones.

Table 2.5 The description of features that were collected from Twitter

Twitter	Attribute	Description
NumOfFollowers	Total number of users following me
NumOfOtherRT	#retweets of my tweets by others *
LikesGivenToMe	#likes my tweets received
NumOfTweets	Total number of tweets
NumOfFollowing	Total number of users I follow
LikesGivenToOthers	Number of tweets that I like
NumOfTxt	#tweets with only text *
NumOfUrl	# tweets that contain URLs *
NumOfMyRT	#tweets that I re-tweet *
TweetsPerDay	#tweets divided by lifetime (days)
NumOfUserMent	#users mentioned tweets *
NumOfHashTag	#hashtags referenced in tweets *

* The measure is computed over the recent 150 tweets (Vilenchik 2020)

Table 2.6 The description of features that were collected from Steam (Vilenchik 2020)

Steam | Attribute	Description
NumOfFriends	Number of friends
CommentsCount	# comments on user's profile
Groups	Number of Groups
ReviewsCount	Number of reviews created by user
Games	Number of Games owned by user
hoursOnRecord	Number of hours the user played
SteamXP	Experience (calculated by Steam)
SummaryLen	Length of user's summary
ScreenshotsCount	# screenshots created by user
Badges	Number of Badges owned by user
SummaryURL	does the user publish URL in his summary (binary)

Table 2.7 The percentage of explained variance per PC

	PC1	PC2	PC3	PC4	PC5	p
Twitter	18.15%	16.2%	13%	10%	(8%)	12
Instagram	29%	19.1%	10.15%	(9%)	(8%)	11
LinkedIn	25.1%	11.7%	10.3%	7.1%	6.8%	15
Steam	27%	13.6%	10.5%	9.7%	(8%)	11
YouTube	29%	19%	12%	(8%)	(7%)	11
Flickr	26.4%	19.7%	14.3%	12.1%	(8.7%)	9

Parenthesized numbers correspond to variance below $1/p$. The last column gives the number of features, p. (Vilenchik 2020)

The second PC v_2 in Twitter is purely Activity. The negatively signed features in v_2 are indicators of robot/spam behaviour: *NumOfUrl* and *NumOfHashtag*. Indeed, the main way of spamming in Twitter is by hashtags (how? Simply include a trending hashtag in your tweet and anyone who clicks the trending topic will see your ad, for free) and URLs, which appear in shortened form on Twitter and make it impossible to know where the URL is leading. The most significant positively singed feature in v_2 is *NumOfTxt*, the number of messages containing only text, and it is typical of benign behaviour. Therefore we may identify v_2 with a quality of Spam Activity. Indeed, v_2 was used as a linear model for spam detection on the set of accounts collected in (Gulec and Khan 2014) with 95% precision and recall rate.

v_3 in Twitter is also purely Activity but its support is dominated by other activity features. The main features of v_3 others, and the number of other users mentioning. These attributes measure the extent to which a user is a content provider. In addition, the feature of retweets from other users appears in an opposite sign to the former, which excludes content providers that don't generate content but just share it. The top accounts in v_3-measure in the sample include news provider *littlebytesnews*, video gaming support *XboxSupport*, and an American teen content provider *ChelseaaMusic*.

Table 2.8 Loadings of the top three PCs for Twitter data

Twitter	PC_1	PC_2	PC_3
NumOfFollowers	0.38	0	0.06
NumOfOtherRT	0.65	0.041	0.09
LikesGivenToMe	0.65	0.04	−0.09
NumOfTweets	0.02	0.32	0.39
NumOfFollowing	0.07	0.06	0.22
LikesGivenToOthers	−0.01	0.31	0.13
NumOfTxt	−0.04	0.55	−0.08
NumOfUrl	0.05	−0.45	0.35
NumOfMyRT	−0.06	0.22	−0.32
TweetsPerDay	0.01	0.35	0.4
NumOfUserMention	0.02	0.16	0.41
NumOfHashtag	0.05	−0.26	0.43

Table 2.9 Loadings of the top three PCs for Steam data

Steam	PC_1	PC_2	PC_3
NumOfFriends	0.33	0.32	0.3
CommentsCount	0.24	0	0.32
Groups	0.31	0.20	−00.3
ReviewsCount	0.25	0.08	−0.55
Games	0.37	−0.23	−0.24
HoursOnRecord	0.16	0.34	0.4
SteamXP	0.42	−0.37	0.2
SummaryLen	0.17	0.44	−0.09
ScreenshotsCount	0.21	0.12	−0.46
Badges	0.46	−0.37	0.1
SummaryURL	0.22	0.44	−0.07

Turning to Steam, and looking at Tables 2.9 and 2.11 we see that none of the four PCs are pure. In other words, PCA cannot be used to characterize users in Steam, or at least, no easy interpretation of the PCs is at hand. For example, looking at users with large projection on v_1 reveals both heavy gamers that have a narrow social circle and low feedback (e.g. a user that played 230 h in the past 2 weeks, earned 4000 badges, but has only 421 friends and received 164 comments on his profile) and light gamers that have a wide social circle and high feedback (e.g. a user that played merely 32 h in the past 2 weeks, but has 1677 friends and received 2300 comments on his profile), and the spectrum in between.

Table 2.10 The values of α (Feedback) and β (Activity) were computed for each PC using Eq. (2.1)

Twitter	α	α avg	β	β avg
PC1	0.98	0.93 ± 0.09	0.02	0.07 ± 0.08
PC2	0.013	0.014 ± 0.08	0.997	0.96 ± 0.08
PC3	0.02	0.02 ± 0.01	0.98	0.98 ± 0.01
PC4	0.006	0.006 ± 0.008	0.994	0.99 ± 0.02
Percentiles	0.025	0.25	0.75	0.975
α	0.022	0.121	0.351	0.627
β	0.379	0.648	0.882	0.978
Instagram	α	α avg	β	β avg
PC1	0.996	0.99 ± 0.01	0.004	0.01 ± 0.03
PC2	2e-05	0.001 ± 0.002	0.99998	0.99 ± 0.003
PC3	0.003	0.01 ± 0.02	0.997	0.99 ± 0.02
YouTube	α	α avg	β	β avg
PC1	0.92		0.08	
PC2	0.003		0.997	
PC3	0.35		0.65	
Percentiles	0.025	0.25	0.75	0.975
α	0.055	0.213	0.490	0.751
β	0.235	0.506	0.783	0.943

The average is taken over 100 random subsamples each of size 5000–10,000 users (depending on the social network). The percentiles were empirically computed over a sample of 10,000 random unit vectors (Vilenchik 2020)

Semantic Shattering

In the previous section we have seen that in Twitter one can easily assign semantic meaning to the PCs and use them to characterize users. Similarly, in Instagram and in YouTube, the top PCs may all be easily interpreted (Table 2.10). On the other hand, in Steam, Flickr and LinkedIn the majority of PCs cannot be interpreted as their energy is spread over Activity and Feedback (Table 2.11). Is this a random phenomenon? Or is there some underlying principal that is responsible for this separation of platforms. In this section we examine this phenomenon in detail.

The key observation to understand what is going on is summarized in the following lemma, which in simple words says that the scores (projections) of a data set X on two different PCs are not correlated. Recall that we use X for the $n \times p$ data matrix (n is the number of samples and p the number of features) and $\hat{\Sigma} = \frac{1}{n} X^T X$

for the sample covariance matrix. We use bold font for vectors, and consider them as column vectors.

Lemma 2.1 Let v_i, v_j be two PCs of $\hat{\Sigma}$ with $i \neq j$. The scores $y_i = Xv_i$ and $y_j = Xv_j$ satisfy $y_i^T y_j = 0$, i.e. they are uncorrelated.

Table 2.11 The values of α (Feedback) and β (Activity) were computed for each PC using Eq. (2.1)

LinkedIn	α	α avg	β	β avg
PC1	0.32	0.36 ± 0.16	0.68	0.64 ± 0.17
PC2	0.6	0.42 ± 0.23	0.4	0.57 ± 0.23
PC3	0.09	0.11 ± 0.09	0.91	0.92 ± 0.08
PC4	0.08	0.08 ± 0.07	0.92	0.9 ± 0.1
PC5	0.08	0.09 ± 0.08	0.92	0.91 ± 0.09
Percentiles	0.05	0.25	0.75	0.94
α	0.092	0.2	0.44	0.6
β	0.387	0.638	0.852	0.897
Steam	α	α avg	β	β avg
PC1	0.16	0.18 ± 0.03	0.84	0.81 ± 0.03
PC2	0.1	0.11 ± 0.03	0.9	0.88 ± 0.03
PC3	0.2	0.15 ± 0.06	0.8	0.85 ± 0.06
PC4	0.11	0.13 ± 0.05	0.89	0.88 ± 0.04
Percentiles	0.025	0.25	0.75	0.975
α	0.005	0.062	0.264	0.550
β	0.442	0.734	0.938	0.994
Flickr	α	α avg	β	β avg
PC1	0.58		0.42	
PC2	0.999		0.001	
PC3	0.09		0.91	
PC4	0.025		0.975	
Percentiles	0.05	0.25	0.75	0.95
α	0.11	0.28	0.60	0.81
β	0.19	0.40	0.72	0.89

The average is taken over 100 random subsamples each of size 2000–10,000 users (depending on the social network). The percentiles were empirically computed over a sample of 10,000 random unit vectors (Vilenchik 2020)

Proof The proof follows immediately from definitions.

$$y_i^T\, y_j = \left(Xv_i\right)^T \left(Xv_j\right) = \left(v_i^T X^T\right)\left(Xv_j\right) = v_i^T\left(X^T X\right)v_j = nv_i^T \hat{\Sigma} v_j.$$

Since v_j is an eigenvector of $\hat{\Sigma}$ we can substitute $\hat{\Sigma}v_j$ with $\lambda_j v_j$ and obtain:

$$nv_i^T \hat{\Sigma} v_j = nv_i^T \lambda_j v_j = n\lambda_j v_i^T v_j = 0. \tag{2.2}$$

The last equality is due to the orthonormality of the eigenvectors. Figure 2.1 illustrates how no-correlation, given by Lemma 2.1, looks like in the Twitter dataset.

Why should we care about this property? Well, if one PC points in the direction of Activity, and another in the direction of Popularity, then Lemma 2.1 may entail that Activity and Popularity are not correlated. This is a very counter-intuitive

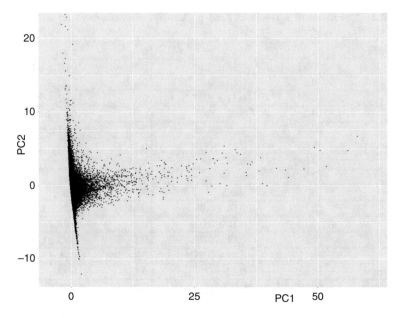

Fig. 2.1 A projection of 285,000 Twitter users on two PCs: v_1 (Feedback) and v_2 (Activity)

conclusion. A healthy life-cycle in online social platforms consists of three fundamental elements, *producing content*: posting opinions, questions, answers, photos,
videos; *consuming content*: viewing videos, reading posts; *giving feedback*: liking,
retweeting, sharing. Producing enables consuming, consuming leads to feedback,
which in turn encourages producing.

Such counter-intuitive observations were already made, for example the "Million
Followers Fallacy", a term coined by (Avnit 2009), who pointed to anecdotal evidence that some users follow others simply because it is polite to follow someone
who is following you. As a result, the OSN contains supposedly-central users with
a huge amount of followers who nobody reads their posted content. Cha et al. (2010)
confirmed the "Million Followers Fallacy" in Twitter by showing that a user's number of followers and his influence (measured as the ability to spread popular news
topics) were not correlated. Other works noted as well that various statistics that one
would naturally assume to be good indicators of one's influence and centrality were
found to be not so, e.g. (Green 2008; Trusov et al. 2010).

The term *semantic shattering*, coined in Vilenchik (2020), generalizes
Lemma 2.1 and provides a wide multidimensional framework to detect semantic
inconsistencies in the data, such as the one pointed by the Million Followers
Fallacy. More formally, suppose that data is collected over m features $f_1, \ldots, f_m$
and that there are r qualities or aspects of interest $Q_1, \ldots, Q_r$, which are defined
using these features. We say that a dataset D *semantically shatters* if there exist
two qualities Q_i and Q_j which are not correlated. For example, in the Million
Followers Fallacy setting we have f_1 the number of followers, f_2 the number of

re-tweets of one's tweets, Q_1 is the quality of Popularity measured using f_1 and Q_2 is the aspect of Influence measured using f_2.

Clearly there are qualities that naturally do not correlate, such as age and the average number of a's in ones posts. Semantic shattering is interesting when qualities that one naturally assumes to be correlated – shatter, such as Popularity and Influence, or Activity and Feedback as found in (Vilenchik 2018).

Theorem 2.2 sets up the formal framework to measure correlation between qualities. It contains three easily-computable sufficient conditions for a dataset D to exhibit semantic shattering. For a vector v, we use the notation $supp(v)$ for the support of v, namely $supp(v) = \{r : v[r] \neq 0\}$. Abusing notation, we use Q_i also for the set of features that define this quality.

Theorem 2.2 (Vilenchik 2020) *Let $v_1, ..., v_p$ be the PCs of the covariance matrix of a p-dimensional dataset D. Let $Q_1, ..., Q_p$ be the qualities spanned by the p features. If there exist two qualities Q_s, Q_t that satisfy the following conditions, then D exhibits semantic shattering.*

1. $Q_s \cap Q_t = \varnothing$
2. Let $A = \left\{v_{i_1}, ..., v_{i_a}\right\}$ be the set of PCs that satisfy $supp(v) \subseteq Q_s$ and let $B = \left\{v_{j_1}, ..., v_{j_b}\right\}$ be those that satisfy $supp(v) \subseteq Q_t$. Furthermore, $A \neq \varnothing$ and $B \neq \varnothing$.
3. *For every $k = 1, ..., p$, either $v_k \in A \cup B$ or $supp(v_k) \cap (Q_s \cup Q_t) = \varnothing$.*

Proof The third condition implies that all the information in the dataset regarding qualities Q_s, Q_t is in the vector space spanned by $A \cup B$. Therefore vectors $v_k \notin A \cup B$ may be ignored. The first and second conditions imply that the vector space spanned by A contains all the information about Q_s and similarly the vector space spanned by B contains Q_t. Lemma 2.1 applied to all the pairs $v_i \in A$ and $v_j \in B$ characterizes the manner in which Q_s and Q_t are uncorrelated.

Note that Theorem 2.2 implicitly assumes that every feature is relevant to at most one quality.

Relaxing Theorem 2

As we have already pointed out in previous sections, the setting of Theorem 2.2 is too "clean" to be relevant for real data. A major obstacle is that every PC v will typically satisfy $supp(v) = \{1, ..., p\}$, yet because numerical reasons, making it impossible to meet the condition of the theorem. Therefore, in order for the framework to be useful it needs to be relaxed. The relaxation suggested in (Vilenchik 2020) is the following. For simplicity of presentation we focus on two qualities Q_1, Q_2. The extension to the general case is straightforward.

Recall Eq. (2.1) that defined For every PC v_i its *energy* in the direction of Q_1 and Q_2. The total energy of every v_i is 1 as it is a unit vector, therefore $\alpha_i + \beta_i = 1$. The conditions of Theorem 2.2 are equivalent in this case to requiring that for every PC v_i, either $\alpha_i = 1$ or $\beta_i = 1$. As in previous sections, this requirement may be relaxed by demanding that $\alpha_i \geq x_{0.975}^{\alpha}$ and $\beta_i \leq x_{0.025}^{\beta}$ or vice versa or vice versa, where $x_{0.975}^{\alpha}$ is the 97.5%-percentile of the α value had the vector v_i been a random p-dimensional vector on the unit sphere. In the same manner all the other percentiles are defined. Recall that this property was named the (α, β)-*separation property*. The choice of 0.975 and 0.025 is somewhat arbitrary, and in general the closer the percentiles are to 1 or 0 the closer we are to the setting of Theorem 2.2.

The second issue is PCs that explain incidental variance (what is informally called noise). Ignoring such PCs is a common practice, known as the Guttman-Kaiser (GK) criterion (Yeomans and Golder 1982), where a PC is considered informative only if it explains more than $1/p$-fraction of the variance. However even among the PCs that pass the GK-criterion, some may fail to satisfy the (α, β)− separation property. If they are border-line with respect to the GK-criterion and their α and β values fall in the inter-quartile regime, the will be classified as Neutral as well, since intuitively their energy is spread between Q_1 and Q_2 as one would expect from a random vector. To summarize the PC classification, a PC v_i is:

- *Purely Q_1* if $\alpha_i > x_{0.975}^{\alpha}$ and $\beta i < x_{0.025}^{\beta}$
- *Purely Q_2* if $\beta_i > x_{0.975}^{\beta}$ and $\alpha_i < x_{0.025}^{\alpha}$
- *Neutral* if $\alpha_i \in \left[x_{0.25}^{\alpha}, x_{0.75}^{\alpha} \right]$ and $\beta_i \in \left[x_{0.25}^{\beta}, x_{0.75}^{\beta} \right]$ and v_i explains roughly 1/p-fraction of the variance.
- *Mixed* in all other cases.

A dataset D exhibits semantic shattering if:

- All the PCs of its covariance matrix that pass the GK-criterion are either purely Q_1, purely Q_2, or Neutral.
- There exists at least one pure PC for every quality.

Otherwise, i.e. if there exists a mixed PC or at least one of the pure types is missing, then we declare that *the framework was unable to detect semantic shattering*.

Shattering in Online Social Media

After surveying in detail the semantic shattering framework, let us see how it applies to each of the six online platforms that we discuss above.

Table 2.10 summarizes the statistics of the α and β values of the PCs that passed the GK-criterion for Twitter, Instagram and YouTube. We conclude that in Twitter v_1 is purely Feedback, v_2, v_3 and v_4 are purely Activity. There are no Mixed or Neutral PCs. Hence we confirm semantic shattering. Similarly, in Instagram, v_1 is purely Feedback and v_2 and v_3 are purely Activity. No Neutral or Mixed PCs, and again we confirm semantic shattering. In YouTube v_1 is purely Feedback and v_2 purely Activity. v_3 explains 12% of the variance, slightly more than $1/p = 1/11 \approx 9\%$, and both its α, β-values fall in the inter-quartile regime. Therefore, we classify it as Neutral, and confirm shattering.

Looking at Table 2.11 we see that in LinkedIn, Steam and Flickr the leading PC is Mixed, namely neither pure Activity nor pure Feedback. This is a dead-end in terms of the semantic shattering framework hence shattering cannot be confirmed. Nevertheless, we see that in LinkedIn v_2 is purely Feedback and v_3, v_4, v_5 are purely Activity. In Flickr v_2 is purely Feedback and v_3, v_4 are purely Activity. Therefore, these pure PCs may be used for user characterization, as discussed in the previous section. In Steam, as we've already mentioned, all fours PCs are Mixed.

Discussion

In this chapter we have seen how PCA may be used for the task of user characterization in on-line platforms, may it be an e-learning platform for practicing mathematics, or an on-line social network. The main advantage of using PCA is that labeled data is not required (unsupervised learning). Labeled data is especially non-trivial to obtain in e-learning platforms where concepts such as "level of engagement" easily evade standards and definitions.

PCA may be used in a static fashion, in which case the PCs are assigned semantic meaning according to the sum of prominent features in their support. Then, users are characterized by the size of their projection on each PC. Sometimes there is no natural interpretation for a PC, in which case it does not take part in this scheme. PCA may be used also in a dynamic fashion. In this case, a time-series is generated for every user by projecting the data on a certain PC in a given frequency. The characterization of the user is then derived from the trends of the series: monotone up? fixed? variable? The underlying working hypothesis is that the more diverse the geometry of the time-series the more diverse are the patterns of usage of the system. And diverse usage leads to more meaningful engagement and in particular a prolonged dwell in the system.

Finally, we have seen a somewhat paradoxical phenomenon: when the PCs are pure, namely amicable for interpretation, then semantic shattering lurks. We have seen a curious partition to platforms where semantic shattering occurs and others where it doesn't. We now try to surmise what stands behind this partition. The two groups of platforms are also separated in their niche-level and in the typical amount of effort it takes to produce content in that network. Twitter, Instagram and YouTube are in some sense "multipurpose" or "generic" OSNs in which content is "virtual" and easily produced. LinkedIn, Flickr and Steam are more thematic-niche and the content shared by a user reflects activities that require considerably larger effort: gaming, semi-professional photography and one's career and education. The niche level and the amount of effort may control the level of *commitment* that the users feel towards each other. Indeed, sociologists typically use the concept of commitment when they are trying to account for the fact that people engage in a consistent manner (Becker 1960).

Taking a closer look at the types of statistics that were collected, and led to the semantic shattering in some platforms, we see that they all fall under what one may call *simple statistics*. As there is no standard definition for this term, it may be defined by example and common sense. As a rule of thumb, simple statistics involve straightforward bookkeeping and counting. For example, the number of users following a certain user on Twitter, the number of posts per day of a certain user on Facebook, the number of views of one's videos on YouTube, the number of badges a Steam user owns, the average length of one's posts, etc. What may not be considered as simple statistics? As a rule of thumb – latent features. For example, statistics that involve text analysis such as the dominant sentiment in one's tweets (happy, sad, complaining), sociolinguistic features (Macaulay 2007) such as what emoji does one use. More complicated statistics require human evaluation such as estimating how provocative is a certain picture or statistics derived from questioners about one's experience when using the platform, e.g. (Bayer et al. 2016).

One important take home message that comes out of this discussion, is that using simple statistics is indeed attractive, the features are easy to compute, and in many cases they lead to a useful user characterization scheme. However, if one measures users using only simple statistics, then Simpson's-type paradoxes may appear in the form of semantic shattering. Indeed, if one looked on sub-communities of Twitter, say a group of extreme mountain climbing aficionados that post pictures from their travels, then one would probably won't see a decoupling between Activity and Feedback. But, to identify such a community one needs to use features which are not simple statistics.

References

Anderson, T. W. (1962). *An introduction to multivariate statistical analysis.* Hoboken: Wiley-Interscience.

Avnit, A. (2009). The million followers fallacy. http://blog.pravdam.com/the-million-followers-fallacy-guest-post-by-adi-avnit.

Bayer, J. B., Ellison, N. B., Schoenebeck, S. Y., and Falk, E. B. (2016). Sharing the small moments: Ephemeral social interaction on snapchat. *Information, Communication & Society* 19: 956–977.

Becker, H. S. (1960). Notes on the concept of commitment. *American Journal of Sociology* 66 (1): 32–40.

Canali, C., Casolari, S., and Lancellotti, R. (2012). A quantitative methodology based on component analysis to identify key users in social networks. *International Journal of Social Network Mining* 1: 27–50.

Cha, M., Haddadi, H., Benevenuto, F., and Gummadi, K. P. (2010). Measuring user influence in Twitter: The million follower fallacy. *Proceedings of the fourth International AAAI conference on weblogs and social media (ICWSM)*. https://www.aaai.org/Library/ICWSM/icwsm10contents.php.

Cocea, M. and Weibelzahl, S. (2007). Cross-system validation of engagement prediction from log files. In: *European conference on technology enhanced learning.* https://link.springer.com/conference/ectel.

Crossley, S., Paquette, L., Dascalu, M., McNamara, D. S., and Baker, R. S. (2016). Combining click-stream data with NLP tools to better understand MOOC completion, *Proceedings of the sixth international conference on learning analytics & knowledge*. https://dl.acm.org/doi/proceedings/10.1145/2883851.

Eirinaki, M., Monga, S. P. S., and Sundaram, S. (2012). Identification of influential social networkers. *International Journal of Web Based Communities* 8: 136–158.

Freeman, L. (1978). Centrality in social networks conceptual clarification. *Social Networks* 1: 215–239.

Green, H. (2008). Google: Harnessing the power of cliques. *BusinessWeek* 50, October 6.

Gulec, A. and Khan, Y. (2014). Feature selection techniques for spam detection on Twitter. *Technical report, Electronic Commerce Technologies (CSI 5389) Project Report School of EE-CS,* University of Ottawa.

Hershcovits, H., Vilenchik, D. and Gal, K. (2020). Modeling engagement in self-directed learning systems using principal component analysis. *IEEE Transactions on Learning Technologies* 13: 164–171.

Katz, E. and Lazarsfeld, P. F. (1957). Personal influence: The part played by people in the flow of mass communication. *American Sociological Review* 21: 61–78.

Lloyd, N., Heffernan, N. T., and Ruiz, C. (2007). Predicting student engagement in intelligent tutoring systems using teacher expert knowledge. *The educational data mining workshop held at the 13th conference on artificial intelligence in education.* Los Angeles, California.

Lykourentzou, I., Giannoukos, I., Nikolopoulos, V., Pardis, G., and Loumos, V. (2009). Dropout prediction in e-learning courses through the combination of machine learning techniques. *Computers & Education* 53: 950–965.

Macaulay, R. (2007). *Talk that counts: Age, gender, and social class differences in discourse.* Oxford: Oxford University Press.

Mislove, A., Marcon, M., Gummadi, K. P., Druschel, P., and Bhattacharjee, B. (2007). Measurement and analysis of online social networks. *Proceedings of the 7th ACM SIGCOMM conference on internet measurement.* https://dl.acm.org/doi/proceedings/10.1145/1298306.

Muirhead, R. (2005). *Aspects of multivariate statistical theory.* Hoboken: Wiley-Interscience.

Pearce, J. (2005). Engaging the learner: How can the flow experience support e-learning? *e-Learn: World conference on e-learning in corporate, government, healthcare, and higher education.* Association for the Advancement of Computing in Education (AACE).

Pennacchiotti, M. and Popescu, A.-M. (2011). A machine learning approach to Twitter user classification. *Proceedings of the fifth International AAAI conference on weblogs and social media (ICWSM).* https://aaai.org/Conferences/ICWSM/icwsm11.php.

Preotiuc-Pietro, D., Lampos, V., and Aletras, N. (2015). An analysis of the user occupational class through twitter content. *Proceedings of the 53rd annual meeting of the Association for Computational Linguistics (ACL).* https://www.aclweb.org/anthology/volumes/P15-2/.

Ramesh, A., Goldwasser, D., Huang, B., Daumé III, H., and Getoor, L. (2013). Modeling learner engagement in MOOCs using probabilistic soft logic. *NIPS workshop on data driven education.* http://www.wikicfp.com/cfp/servlet/event.showcfp?eventid=32328&copyownerid=54396.

Ramesh, A., Goldwasser, D., Huang, B., Daume III, H., and Getoor, L. (2014). Learning latent engagement patterns of students in online courses. *Twenty-eighth AAAI conference on artificial intelligence.* https://www.aaai.org/Library/AAAI/aaai14contents.php.

Rao, D., Yarowsky, D., Shreevats, A., and Gupta, M. (2010). Classifying latent user attributes in twitter. *Proceedings of the 2nd international workshop on search and mining user-generated contents.* https://dl.acm.org/doi/proceedings/10.1145/1871985

Rodrıguez-Ardura, I. and Meseguer-Artola, A. (2017). Flow in e-learning: What drives it and why it matters. *British Journal of Educational Technology* 48: 899–915.

Trusov, M., Bodapati, A., and Bucklin, R. (2010). Determining influential users in internet social networks. *Journal of Marketing Research* 47: 643–658.

Vilenchik, D. (2018). The million tweets fallacy: Activity and feedback are uncorrelated. *Proceedings of the twelfth International AAAI conference on weblogs and social media (ICWSM)*. https://www.aaai.org/Library/ICWSM/icwsm18contents.php.

Vilenchik, D. (2020). Simple statistics are sometime too simple: A case study in social media data. *IEEE Transactions on Knowledge Data Engineering* 32: 402–408.

Vilenchik, D., Yichye, B., and Abutbul, M. (2019). To interpret or not to interpret PCA? This is our question. *Proceedings of the thirteenth International AAAI Conference on Web and Social Media (ICWSN)*. https://www.aaai.org/ojs/index.php/ICWSM/article/view/3265.

Yang, D., Sinha, T., Adamson, D., and Rosé, C. P. (2013). Turn on, tune in, drop out: Anticipating student dropouts in massive open online courses. *Proceedings of the 2013 NIPS data-driven education workshop*. http://www.wikicfp.com/cfp/servlet/event.showcfp?eventid=32328&copyownerid=54396.

Yeomans, K. A. and Golder, P. A. (1982). The Guttman-Kaiser criterion as a predictor of the number of common factors. *The Statistician* 31: 221–229.

Chapter 3
Argumentation Is Elementary: The Case for Teaching Argumentation in Elementary Mathematics Classrooms

Cathy Marks Krpan and Gurpreet Sahmbi

Introduction

Navigating the information age successfully requires that consumers of data assess and critique the validity of its content and in doing so, create a logical argument that supports or refutes its truth. Furthermore, the skill sets and qualifications expected of the workforce no longer consist of following simple routines, but necessitate problem solving and the understanding of complex systems through constructing, describing, explaining, manipulating, and predicting (English et al. 2013). These skills are central to mathematics, and we believe that, more specifically, the development of strong argumentation and reasoning skills in mathematics enables learners to navigate the complexities and the increasing volume of information they encounter. In this chapter, we contend that argumentation needs to play a more significant role in elementary mathematics and that its implementation can enable learners to develop key skill sets that are essential for deep mathematical thinking and reasoning skills.

This chapter offers a rationale for the use of argumentation in elementary mathematics classrooms. The authors present an overview of the term argumentation (and its related mathematical counterpart, proof), its implementation in mathematics classrooms, including benefits and examples of effective use, and challenges teachers face. We contend that argumentation as a mathematical process supports students not just in the development of deeper content knowledge, but in developing the twenty-first century skills of critical discourse, problem solving, and logic building that are necessary for success.

C. M. Krpan (✉) · G. Sahmbi
University of Toronto, Toronto, ON, Canada
e-mail: cathy.marks.krpan@utoronto.ca; gurpreet.sahmbi@utoronto.ca

© Springer Nature Switzerland AG 2020
S. A. Costa et al. (eds.), *Mathematics (Education) in the Information Age*,
Mathematics in Mind, https://doi.org/10.1007/978-3-030-59177-9_3

Argumentation in Elementary Classrooms

There is much debate in the mathematics community about the appropriate use of the terms "argumentation" and "proof" in mathematics (Balacheff 2002; Cirillo et al. 2015; Weber 2014). As Hanna and de Villiers (2008: 331) note, "Some researchers see mathematical proof as distinct from argumentation, whereas others see argumentation and proof as parts of a continuum rather than as a dichotomy." Because of this confusion, many believe that the mathematics community needs to refine the meaning of these terms in research and curriculum documents (Cai and Cirillo 2014). Balacheff (2002: 2) explains that "Our epistemology of proof is the first deadlock to figure out and to cope with when entering our research field." We use the terms "argumentation" and "proof" in the context of this chapter to refer to mathematical thinking that learners use to prove or disprove mathematical claims. This includes mathematical thinking that serves as a precursor to developing a formal proof, in addition to student argumentation that conclusively proves and generalizes mathematics. Both terms, "proof" and "argumentation" will be used interchangeably in this chapter.

In the mathematical community, people hold different ideas about the role of proof and its key elements (Hanna 2000). The notion of argumentation in mathematics can include the examination of a mathematical conjecture or claim, and it can include the development of a logical, sound argument to demonstrate whether it is true or false (Marks Krpan and Sahmbi 2019). Argumentation in mathematics involves creating claims, providing evidence to support those claims, and evaluating evidence to assess their validity (Knudsen et al. 2014). As Solar et al. (2020): 1 describe, "[argumentation] is used to convince oneself and others of the validity of a line of reasoning." Thus, the goal of creating mathematical arguments is to determine the truth of mathematical statements (Knudsen et al. 2014). Some consider argumentation as the key component of inquiry-based classrooms in which students are expected to "propose and defend mathematics ideas and conjectures" (Goos 2004: 259). In essence, argumentation in mathematics helps us to determine why things work (Stylianides et al. 2013).

The development of a sound argument is less about the application of the elements of the proof and more about using the act of proving to deepen one's knowledge about mathematics (Marks Krpan 2018; Marks Krpan and Sahmbi 2020). As Sowder and Harel (1998: 297) point out "The goal [of proving] is to help students define their own conception of what constitutes justification in mathematics." Mathematical argumentation can also support the development of a student's conceptual understanding in mathematics (Rumsey 2012). Hanna (2000: 7) describes the role of mathematical proofs as promoting mathematical understanding, reflecting that, "It became clear to me that a proof, valid as it might be in terms of formal derivation, actually becomes more convincing and legitimate to a mathematician only when it leads to real mathematical understanding." She notes that proofs should have a prominent place in the mathematics curriculum, pointing out that the key role of educators is to understand the role of proofs in mathematics in order to enhance their implementation (Hanna 2000). Ball et al. (2002: 907) also believe that proof should play a significant role in mathematics stating that "proof is central to mathematics and as such, should be a key component of mathematics education. This emphasis can be justified not only because proof is at the heart of mathematical practice, but also mathematical understanding."

In addition to requiring students to explain their mathematical solutions and/or strategies, mathematical argumentation engages the learner in deepening their thinking as they try to convince others of their reasoning (Francisco and Maher 2005; Marks Krpan 2018). When demonstrating that a conjecture is true or untrue, the learner needs to apply their own mathematical knowledge to support or refute a claim. In many cases, students apply their mathematical knowledge in ways that they did not formally learn in math class (Marks Krpan and Sahmbi 2020). The learner makes connections between the mathematics they know and the argument they are trying to make. It is a process through which students may develop an argument about a mathematical concept and rationalize its truth or untruth through mathematical reasoning (Stylianides et al. 2013). Stylianides (2007a: 291) provides the following definition of proof in elementary classrooms along with three key elements he believes are critical:

> Proof is a mathematical argument, a connected sequence of assertions for or against a mathematical claim, with the following characteristics:
>
> 1. It uses statements accepted by the classroom community (set of accepted statements) that are true and available without further justification;
> 2. It employs forms of reasoning (modes of argumentation) that are valid and known to, or within the conceptual reach of, the classroom community; and
> 3. It is communicated with forms of expression (modes of argument representation) that are appropriate and known to, or within the conceptual reach of, the classroom community.

Criteria like these can serve as a practical and usable guide for teachers as they implement argumentation in their classrooms (Marks Krpan 2018).

Argumentation is a social activity, as the very nature of creating an argument requires that one engages in mathematical discourse with others in order to share and debate different mathematical claims related to proving or disproving mathematical ideas (Marks Krpan 2018). The act of proving requires many skills such as conjecturing, justifying, and organizing mathematical ideas in a logical, clear manner to elucidate one's reasoning to others (Knudsen et al. 2014; Marks Krpan 2018). It is critical that students have opportunities to build off of and respond to each other's ideas (Knudsen et al. 2014). During the sharing process, students may edit and clarify their ideas based on the feedback they receive from others, developing an awareness of their own thinking and that of others (Marks Krpan 2018; Marks Krpan and Sahmbi 2020). As students share mathematical arguments, they must also assess and evaluate the credibility of the arguments of their peers (Marks Krpan and Sahmbi 2020). Ayalon and Even (2014) note that mathematics is intrinsically connected to argumentation as it requires that the learner justify claims and generate conjectures which are critical elements of doing and communicating mathematics.

While there is significant research in the area of proof and proving at the secondary and university levels, there is much less research on argumentation and its implementation in elementary classrooms (Hanna 2000; Stylianides 2007b, 2016; Yackel and Hanna 2001). Though students are often expected to understand the concept of a proof, and produce proofs at secondary and post-secondary levels, students are rarely introduced to this kind of thinking at the elementary level (Stylianides 2007b). The absence of proof-building is often evident in the mathematics resources teachers use. Bieda et al. (2014) found that the average percentage of reasoning-and-proving tasks was 3.7% of the total tasks in elementary textbooks they examined. Thus, it is common for students to encounter proof for the first time in secondary or post-secondary schooling, where they will have had little to no background in formalizing their understanding or skills prior to advanced mathematics courses.

Proofs are often associated with strong mathematical rigor and formality and as a result, proofs are usually introduced later in school, and/or university (Francisco and Maher 2005). Sowder and Harel (1998: 674) note that "To delay exposure to reason giving until the secondary-school geometry course, and to then expect an instant appreciation for more sophisticated mathematical justifications is an unreasonable expectation." Consequently, Ball et al. (2002: 908) strongly advocate for "a culture of argumentation in the mathematics classroom from the primary classrooms all the way up through to college."

Stylianides (2016) acknowledges how the view that proofs should only be taught at the secondary and post-secondary levels in education has changed. He points out that in 1989, the National Council of Teachers of Mathematics' (NCTM) Curriculum and Evaluation Standards, which serves as a framework for curriculum development in mathematics, suggests that proofs be explored solely by high-school students who are college bound, while the updated version of this document, published in 2000 (NCTM 2000) stresses that reasoning and proof are "fundamental" elements for all learners of mathematics of all ages. Francisco and Maher's (2005) longitudinal study on problem solving further supports this view, finding that students can readily engage in "proof making" in the early years of education. Yackel (2001: 10) also believes that young learners can fully participate in development of mathematical arguments stating, "I have documented that children as young as second grade engage in sophisticated forms of explanation and justification and that their understanding of explanation advances as the school year progresses."

The value of introducing proofs in elementary classrooms is underscored by Hanna and de Villiers' (2008: 329) research in which they explore the concept of "developmental proofs" which they describe as a "precursor to formal proofs that grows in sophistication as the learner matures towards more coherent conceptions." They stress that proof and proving in elementary classrooms have the potential to facilitate the development of more formal proofs in mathematics. If an emphasis on mathematical argumentation can be fostered in the elementary grades, encouraging students to justify and think deeply about mathematics, students can be prepared for writing proofs in secondary mathematics (Cervantes-Barraza et al. 2020; Hoffman et al. 2009). Rumsey (2013: 121) stresses that "We need information about early reasoning and proof if we are to help students transition along the continuum from early mathematics and informal proof toward grade levels requiring formal proof." Yet, in spite of current literature, argumentation is still considered an area of mathematics that is more conducive to secondary and/or university mathematics (Bieda et al. 2014; Stylianides et al. 2013; Stylianides 2007b). This view often limits its implementation in elementary classrooms (Marks Krpan and Sahmbi 2019; Stylianides 2016).

Implementation of Argumentation

Though influential recommendations such as the NCTM's Curriculum and Evaluation Standards (2000) have encouraged the use of argumentation and proof-building in elementary mathematics, implementation has been and is currently highly variable. Research suggests that successful implementation of argumentation in mathematics classrooms hinges on the ability of a teacher to cultivate and sustain a classroom culture that is conducive to proving mathematical ideas. More specifically, teachers must foster an environment that supports student collaboration and discussion as it pertains to mathematics (Civil and Hunter 2015).

Creating a Classroom Culture for Proving

The classroom environment plays a critical role in facilitating mathematical discussions and the sharing of mathematical ideas (Cervantes-Barraza and Cabañas-Sánchez 2018; Marks Krpan 2013). In order for students and teachers to successfully engage in the act of proving and the development of argumentations, there must be opportunities for student collaboration and exploration (Civil and Hunter 2015; Marks Krpan 2018; Marks Kpran and Sahmbi 2020; Rumsey and Langrall 2016). In addition, the classroom culture needs to foster and support student risk taking (Marks Krpan 2018). For example, Makar et al. (2015) note that for argumentation to successfully take place, students need to be encouraged to think without worrying about having a correct or complete answer related to their argumentations. This opportunity to think aloud can encourage students to take what the authors describe as "intercultural risks to share emergent and incomplete ideas" (Makar et al. 2015: 1116).

Schwarz et al. (2010: 108) remind us that "for many influential researchers in mathematics education, argumentation is seen as the way meaning-making and understanding develop[s] in classroom discussions." In order to deepen their reasoning skills and make meaning, students need opportunities where they can defend their mathematical claims and respond to the argumentation of their peers (Goos 2004). It is through whole class and small group discussions that students engage in mathematical argumentation and justify their thinking to others (Marks Krpan 2018;

Marks Krpan and Sahmbi 2020; Whitenack and Yackel 2002). In addition, these mathematical discussions are critical to the argumentation process as they enable students to learn from each other (Marks Krpan 2018; Marks Krpan and Sahmbi 2020).

Beyond deepening their understanding of mathematics, class discussions can enable students to discover new ways of reasoning mathematically (Civil and Hunter 2015; Marks Krpan and Sahmbi 2020; Rumsey et al. 2019; Whitenack and Yackel 2002). New learning can originate from individual student argumentations which in turn can inform whole class discussions. (Civil and Hunter 2015; Marks Krpan 2001; Marks Krpan and Sahmbi 2020; Whitenack and Yackel 2002). Furthermore, students can improve their mathematical content and concept knowledge as they explore ideas through mathematical discussions (Mercer 2010; Solomon 2009). By encouraging students to share their insights with each other, teachers provide opportunities for students to rehearse their thinking and link ideas together (Zolkower and Shreyar 2007). These argumentation discussions enable teachers to have access to the diversity of student reasoning and argumentation skills (Reid et al. 2011).

The act of proving engages the learner in a social process in which they share their mathematical insights and try to convince others of their reasoning (Rumsey 2013; Marks Krpan 2018). Solar et al. (2020: 24) note that "communication strategies, such as giving students opportunities to participate, managing error, and asking deliberate questions have proven to be highly relevant in supporting argumentation among students and incorporating contingencies into classroom discussions," In addition, student engagement (Kazemi and Stipek 2009; Turner and Meyer 2004) and self-directed learning (Francisco and Maher 2011) can improve when learners have opportunities to share, compare and justify their ideas to each other.

Collective argumentation can support and guide collaborative engagement and student discourse in mathematics as students engage in the act of proving (Brown 2017; Cervantes-Barraza et al. 2020; Lin and Tsai 2012). It is characterized by a set of interactions that takes place among students as they convince each other of their mathematical arguments and arrive at a solution within whole class discussions facilitated by the teacher (Krummheuer 2015). Brown (2017: 186) believes that collective argumentation can have a positive impact on learning in primary classrooms. He describes collective argumentation based on five principles: generalizability (students communicate their own thinking about a task), objectivity (ideas can only be rejected by using logical arguments); consistency (ideas make sense); consensus (all students in the group understand, agree and can explain their argumentation); and recontextualization (students present their ideas to the class for discussion and validation). Brown found that implementation of these principals not only increased student engagement but also improved the overall quality of teaching and learning.

The teacher plays a critical role in facilitating and orchestrating mathematical discussions by establishing classroom norms for interaction (Marks Krpan 2018;

McCrone 2005). In a classroom environment that encourages respectful discourse, students can work collaboratively and gain a collective understanding when working on argumentation tasks (Civil and Hunter 2015). Yackel (2004: 6) advocates for classroom discussions in her description of the social norms necessary for the development of argumentation skills:

> Social norms that characterized classroom interactions [...] include that students are expected to develop personally-meaningful solutions to problems, to explain and justify their thinking and solutions, to listen to and attempt to make sense of each others interpretations of and solutions to problems, and to ask questions and raise challenges in situations of misunderstandings and disagreement.

Accountable talk c.an only take place when students understand that they are expected to listen to others, build on each other's ideas, provide explanations and justifications for their ideas, and challenge the thinking of others when necessary (Michaels et al. 2008).Establishing expectations of interactions such as how to actively listen and disagree in a respectful manner can improve student learning and create a supportive learning culture (Marks Krpan 2011, 2013).

Expectations that delineate what collaborative behaviours in mathematics "look like," "sound like," and "feel like," can enhance student interactions (Marks Krpan 2011). Yackel (2001) points out that in mathematics we need to stress the importance of "social norms" (expectations developed in the classroom) and "sociomathematical norms" (related to interactions specific to mathematics) in the context of promoting explanation, justification, and argumentation. She uses the term "group think" to refer to collective argumentation in which students reason interactively through whole group and small group contexts. Yackel (2001: 15) argues that "the reason that group think is so powerful for students' learning is that it emphasizes what most mathematicians and mathematics educators consider to be the essence of mathematics – mathematical reasoning and argumentation."

Another example that stresses the role of expectations for engagement can be found in McCrone's (2005) work. In her study, a grade 5 teacher read a story about three characters who arrived at different solutions to a mathematics problem about money. Students were invited to work in pairs to determine which of the three solutions they agreed with and come up with ways to convince their classmates of their answer. At first, students engaged mostly in parallel discussions when working with their partners, not realizing that commenting on their classmate's work was also part of their role. McCrone (2005) noted the importance of the teacher establishing expectations for engagement and modelling active listening for her students which assisted students in reflecting on their own thinking when listening to classmates describe theirs. By the end of the study, students took on more of a responsibility in understanding the reasoning of others and were more able to make sense of each other's thinking.

Implementation of Argumentation in Elementary Classrooms: Some Examples

There are a variety of teaching approaches that can be used to explore argumentation in the elementary grades (Knudsen et al. 2014; Marks Krpan and Sahmbi 2019). It is critical that teachers have access to teaching strategies that are easy to implement and enable their students to fully engage in the creation of meaningful argumentations (Marks Krpan 2018; Marks Krpan and Sahmbi 2020).

In a recent study by Rumsey and Langrall (2016), teachers taught number sense through argumentation tasks. Students were provided with a claim such as "any number multiplied by 0 will equal a larger number than 100" and were asked to complete "language frames" which included prompts such as "I agree because…" or "I disagree because…" (Rumsey and Langrall 2016: 414).They found that the language frames supported the development of discourse related to argumentation and assisted students in sharing ideas which led to rich whole-class discussions. The authors also noted that organizing small and whole-class discussions were beneficial, as some students preferred to share in smaller group settings.

Journal writing can also be used to assist students in developing argumentation skills. Bostiga et al. (2016) investigated the use of debate journals with grade 3 and 5 students to help them develop and assess mathematical arguments through writing. Like Rumsey and Langrall (2016), they utilized prompts to guide student thinking. The journal prompts included correct or incorrect statements, based on common student misconceptions. For example, one prompt (with images provided) asked, "When asked to shade 4 tenths, Corey shaded 4 rows and David shaded 4 squares. Do you agree with Corey or David?" (Bostiga et al. 2016: 550). Students were invited to explain in writing with which statement they agreed with and why. Teachers discussed the statements with the whole class beforehand and then invited students to solve them on their own and compare their answers with their classmates. Bostiga et al. (2016) found that the debate journals immensely improved both students' mathematical writing and their argumentation skills.

Existing textbook content can be modified to include more argumentation opportunities for students. Rumsey (2013) modified textbook lessons related to properties of multiplication by including more opportunities for argumentation and justification. Modifications to the textbook lessons included adding additional problems and counter-examples, emphasizing the language of argumentation, promoting discussions about number properties and including true/false and open number sentences. Rumsey (2013: 10) found that "even without prior instruction emphasizing argumentation, students were able to bring relevant knowledge regarding arguments to the discussion that could be used to help them transition to more formal proof."

Some researchers have examined collective argumentation tasks as a way to implement argumentation in elementary classrooms. Cervantes-Barraza et al. (2020) investigated a collective argumentation task in which they invited grade 5 students to prove the theorem that opposite angles of a vertex were equal. The structure of the task, like Brown's (2017), was based on a set of principles to guide the learning process. Cervantes-Barraza et al. (2020), based their task on Lin and Tsai's (2012) principles of mathematical proof in which students are encouraged to analyze the problem and find patterns, build a conjecture based on their observations, transform their observations into generalizations and then reflect on the generalization they made and validity of their proof. Cervantes-Barraza et al. (2020) noted that collective argumentation enabled students to make generalizations about opposite angles. They also concluded that collective participation played a critical role in the development of proofs as it fostered interest and curiosity among the students and also provided opportunities for engagement for students who did not usually participate in mathematics discussions.

The authors of this chapter have also explored argumentation tasks with grade 2, 3 4 and 5 students who had not previously engaged in any formal argumentation tasks (Marks Krpan and Sahmbi 2020). Students were presented with number statements that the classroom community already knew to be false. The argumentation task involved providing students with a number statement such as $10 + 10 = 12$ or $6 \times 4 = 25$ (depending on the grade level) and inviting students to argue, using visuals, numeric notation, or written explanations, why it was false. Teachers noted that not only did students apply mathematics concepts and strategies they had learned from past mathematics lessons, but also applied previously learned mathematics content in novel ways. Teachers observed that students modified the use of number lines in their counter arguments from how they were used in the teaching. For example, when disproving the statement $6 \times 4 = 25$ one group of students compared two different number lines side by side while another group showed jumps of 4 on their number line and in addition, noted the accumulative total of each jump on the same number line. This, the teachers noted was different from how they used number lines in their teaching. Even though the students' work did not represent formal mathematical generalizations, their work demonstrated high-level mathematical thinking and included solid counter arguments (Marks Krpan and Sahmbi 2020). Our findings indicate that this teaching approach is an effective precursor to the development of formal proofs and enables learners of all ages to have access to develop critical argumentation skills.

Challenges in Implementation

While there is a plethora of evidence suggesting the benefits of argumentation in elementary classroom (Marks Krpan 2018; Mercer 2010; Rumsey 2012; Solomon 2009), including studies that illustrate examples (Bostiga et al. 2016; Cervantes-Barraza et al. 2020; Rumsey and Langrall 2016), there remains a dearth of implementation in practice. It is evident that teachers play a critical role in weaving argumentation into their practice. Consequently, teacher uptake of this strategy is a determining factor in the pervasiveness and effectiveness of argumentation in the elementary classroom (Stylianides 2016). Studies have shown that teachers feel there are significant challenges that impede their ability to consistently and effectively use argumentation in their mathematics classrooms (Ayalon 2019; Flegas and Charalampos 2013). Most commonly, this centres around the following: (1) lack of teacher knowledge of proof and argumentation (Bieda 2010); (2) difficulty facilitating mathematical discourse (Conner et al. 2014); and (3) limited time and resources (Staples and Newton 2016).

Shulman (1986) discusses the concepts of subject content knowledge and pedagogical content knowledge (PCK) as two areas of teaching that work in tandem to support effective learning experiences for students. Briefly, subject content knowledge is knowledge of the discipline (mathematical content knowledge [MCK]), and PCK is knowledge of how to teach said discipline, or various topics within that discipline. There is extensive research on the importance of both content knowledge and pedagogical content knowledge in mathematics (Ball et al. 2008; Krauss et al. 2008) that suggests both these types of knowledge are critical for effective teaching. For local context, formal proof and argumentation are virtually absent in the elementary curriculum expectations in Ontario (Ontario Ministry of Education 2005a), and only briefly touched upon in the secondary curriculum documents (Ontario Ministry of Education 2005b, 2007). Thus, it is unsurprising that research indicates that teachers tend to be hindered in their implementation of argumentation in classrooms due, in part, to low PCK and MCK as it relates to proof and argumentation (Ayalon 2019; Ayalon and Hershkowitz 2018; Bieda 2010). Indeed, Bieda (2010: 353) notes, "teachers' knowledge of proof and their beliefs about teaching proof may also constrain their ability to teach proof effectively."

Teachers and students, alike, are typically unfamiliar with argumentation tasks or proving in mathematics (Goulding et al. 2002; Knuth 2002; Stylianides et al. 2013). Several studies indicate that low levels of teacher comfortability with the concept of proof is a major hurdle in implementing proof and argumentation in mathematics classrooms (Stylianides 2007a). In a study examining the uptake of logical reasoning and proof in a grade 6 mathematics classroom, Flegas and Charalampos (2013) found that challenges arose for the teacher as they tried to develop effective pedagogical knowledge of how to teach using proof. Notably, the teacher's PCK for proof acted as a barrier to their consistent implementation of this form of mathematics. Even at the secondary level, where proof is more widely understood, Ayalon (2019: 190) argues that "teachers are not adequately trained to identify and enhance argumentation opportunities." and consequently, they "may have difficulties engaging students in constructing and responding to arguments" (p. 192). This suggests limitations in PCK that hinder teachers' ability to compensate for limited resources (discussed later), and effectively engage students in argumentation.

Proof and argumentation as mathematical concepts present additional challenges, as they are existing and central tenets of formal mathematics (Hanna and Jahnke 1996). Thus, they are not simply pedagogical strategies that are used to understand other mathematical concepts but are part of the discipline of mathematics itself. Consequently, teachers must understand this fundamental part of mathematics as well as the mathematical concepts they aim to teach to their students. Indeed, Ayalon and Hershkowitz (2018: 163) state that teachers require knowledge of "the kinds of justifications accepted in mathematics, students' common tendencies and difficulties, and the conditions essential to establish a classroom environment that fosters argumentation." This suggests that teachers are tasked with the traditional issues of MCK and PCK as well as specific understandings of the theoretical basis of proof and argumentation if they are to be effective in implementing proof and argumentation in their classrooms. Hence, it is evident that "teachers' knowledge of proof and their beliefs about proof may also constrain their ability to teach proof effectively" (Bieda 2010: 352).

Fundamental to effective implementation of argumentation in the elementary mathematics classroom is a teacher's comfortability and agility with facilitating mathematical discourse. Though the benefits of promoting mathematical discourse in the classroom have been discussed in the literature for many years (Ball 1991; NCTM 2000), several studies have found that teachers find it challenging to effectively implement this type of pedagogy (Bieda 2010; Stylianides et al. 2013). In a study focused on secondary mathematics teachers' implementation of mathematical argumentation in everyday coursework, Kosko et al. (2014) observed that unfamiliarity with questioning techniques impeded teachers' ability to effectively develop strong mathematical discourse in their classrooms. They noted, "Teachers may not have a clear understanding of what effective questioning strategies look like or how to implement them" and suggested that "more explicit instruction for teachers in how to facilitate mathematical argumentation and discussion" is critical (Kosko et al. 2014: 474). Teachers often struggle to withhold the "correct answer" from students while they work through their arguments (Conner et al. 2014), and the worry that students will stray far from the target concepts results in discomfort. Further, and specific to proof and argumentation, Stylianides (2007b: 17) explains that teachers have the additional challenge of not just facilitating mathematical discussions, but of establishing "socially accepted rules of discourse relevant to proving that are compatible with those of wider society." This alludes to the broader context of proof as a mathematical language with axioms and structure agreed upon by mathematics communities, and the layers of difficulty this presents when teachers are newly engaging in mathematical discourse in this domain.

All these challenges are further compounded by the issue of limited resources available to elementary teachers that explicitly support argumentation tasks (Bieda et al. 2014; McCrory and Stylianides 2014). Bieda et al. (2014) found in their investigation of elementary mathematics textbooks that opportunities to explore mathematical concepts using argumentation were sparse, suggesting a lack of resources readily available for elementary teachers. This finding was corroborated in the same year by researchers who noted that there was a dearth of explicit resources available that focused on argumentation-like tasks or reasoning and proving (McCrory and Stylianides 2014). Consequently, even if teachers wish to use argumentation in elementary mathematics, they often have little outside help on which to rely.

Additionally, while teachers in numerous studies see the conceptual benefits to using argumentation in their classrooms, many noted that the time investment required to actively and effectively use argumentation was a major hindrance in implementation (Bieda 2010; Brodahl and Wathne 2018; Brown 2017). The pressure of curriculum demands in schools imposes restrictions on teachers' time and ability to pursue argumentation in their classrooms. Staples and Newton (2016: 299) explain:

In the context of K–12 schools, with clear content demands and assessment procedures, teachers may have fewer opportunities (real or perceived) to (a) organize lessons that provide opportunities for participation in argumentation, where the purpose is not conceptual development, but deliberate inquiry into the truth of various claims; and (b) take advantage of such opportunities should they arise.

Whether "real or perceived", the issue of time has tangible effects on teacher implementation. Further, teachers find that developing argumentation tasks and framing their classrooms as inquiry spaces where students often misunderstand tasks and hence, can go off-topic, can be time consuming (Brodahl and Wathne 2018). Even in instances where students were successful in using argumentation to engage with primary mathematics, the time for implementation and student discussion was seen as a major challenge (Brown 2017).

Summary

In spite of its many benefits, argumentation is not widely used in elementary classrooms. The reasons for this are complex and cannot be isolated to one factor. Teachers strive to ensure that their students develop key mathematics skills through a variety of tasks and teaching approaches. It is critical that teachers have the opportunity to learn about the role that argumentation can play in supporting student learning in mathematics. The importance of implementing argumentation tasks at the elementary level cannot be overstated. Essential mathematical skills such as communication, justification and reasoning are developed through the act of argumentation and proving. Students in elementary schools are capable of exploring argumentation tasks and engaging in the act of proving. Indeed, there are meaningful ways to infuse argumentation into elementary mathematics programs while supporting all students. However, strong teacher pedagogical knowledge and content knowledge, related to argumentation, are needed in order for teachers to be able to facilitate insightful discussions, and create classroom cultures in which mathematical argumentations can be shared and explored successfully.

Not only will experience with argumentation tasks help students to become better mathematical thinkers, it will also provide them with the foundation necessary to engage in more complex proofs at the secondary and university levels. More importantly, the teaching of argumentation in elementary grades can strengthen students' thinking and reasoning skills which are essential for life-long learning. As students explore their own arguments and those of others, they will develop a deep understanding not only of the mathematical content they are learning, but of complexities of ideas and how others may understand and represent them differently.

Some of the key challenges of implementing argumentation tasks are teachers' unfamiliarity with argumentation and its implementation, and limited time and material resources. We believe that in order to mediate these challenges, argumentation needs to take a prominent place in all elementary mathematical curriculums, including detailed explanations of what argumentation entails and examples of implementation. Moreover, we recommend more professional development for teachers in which they can collaboratively explore the nature of argumentation, its benefits, and how it can be implemented in their mathematics classrooms. In addition, we advocate for continued research in the area of elementary argumentation in order for the mathematical community to gain further insight about its implementation and impact on student learning. Only when these recommendations for the implementation of argumentation in elementary classrooms are put into practice, do we believe that students will be able to have access to the key skillsets of mathematics which are essential to becoming strong, critical thinkers in the information age.

References

Ayalon, M. (2019). Exploring changes in mathematics teachers' envisioning of potential argumentation situations in the classroom. *Teaching and Teacher Education* 85: 190–203.

Ayalon, M. and Even, R. (2014). Factors shaping students' opportunities to engage in argumentative activity. *International Journal of Science and Mathematics Education* 14: 575–601.

Ayalon, M. and Hershkowitz, R. (2018). Mathematics teachers' attention to potential classroom situations of argumentation. *Journal of Mathematical Behavior* 49: 163–173.

Balacheff, N. (2002). The researcher epistemology: a deadlock for educational research on proof. In: *2002 International conference on mathematics-understanding proving and proving to understand* (23–44).

Ball, D. L. (1991). What's all this talk about "discourse"? *The Arithmetic Teacher* 39 (3): 44.

Ball, D. L., Hoyles, C., Jahnke, H. N., and Movshovitz-Hadar, N. (2002). The teaching of proof. *Proceedings of the ICM*, Beijing 2002, 3, 907–922.

Ball, D. L., Thames, M. H., and Phelps, G. (2008). Content knowledge for teaching: What makes it special? *Journal of Teacher Education* 59 (5): 389–407.

Bieda, K. N. (2010). Enacting proof-related tasks in middle school mathematics: Challenges and opportunities. *Journal for Research in Mathematics Education* 41 (4): 351–382.

Bieda, K. N., Ji, X., Drwencke, J., and Picard, A. (2014). Reasoning-and-proving opportunities in elementary mathematics textbooks. *International Journal of Education Research* 64: 71–80.

Bostiga, S. E., Cantin, M. L., Fontana, C. V., and Casa, T. M. (2016). Moving math in the write direction: Reflect and discuss. *Teaching Children Mathematics* 22: 546–554.

Brodahl, C. and Wathne, U. (2018). Imaginary dialogues: In-service teachers' steps towards mathematical argumentation in classroom discourse. *Journal of the International Society for Teacher Education* 22: 30–42.

Brown, R. (2017). Using collective argumentation to engage students in a primary mathematics classroom. *Mathematics Education Research Journal* 29: 183–199.

Cai, J. and Cirillo, M. (2014). What do we know about reasoning and proving? Opportunities and missing opportunities from curriculum analyses. *International Journal of Educational Research* 64: 132–140.

Cervantes-Barraza, J. A., and Cabañas-Sánchez, G. C. (2018). Formal and visual arguments in collective arguments. *Mathematics Education* 30(1): 148–168.

Cervantes-Barraza, J. A., Moreno, A. H., and Rumsey, C. (2020). Promoting mathematical proof from collective argumentation in primary school. *School Science and Mathematics* 120: 4–14.

Cirillo, M., Kosko, K., Newton, J., Staples, M., Weber, K., Bieda, K., and Conner, A. (2015). Conceptions and consequences of what we call argumentation, justification, and proof. In:

Proceedings of the 37th annual meeting of the North American chapter of the psychology of mathematics education (1343–1351).

Civil, M., and Hunter, R. (2015). Participation of non-dominant students in argumentation in the mathematics classroom. *Intercultural Education* 26(4): 296–312.

Conner, A., Singletary, L. M., Smith, R. C., Wagner, P. A., Francisco, R. T. (2014). Teacher support for collective argumentation: A framework for examining how teachers support students' engagement in mathematical activities. *Educational Studies in Mathematics* 86: 401–429.

English, L. D., Hudson, P., and Dawes, L. (2013). Engineering-based problem solving in the middle school: design and construction with simple machines. *Journal of Pre-College Engineering Education Research* 3: 43–55.

Flegas, K. and Charalampos, L. (2013). Exploring logical reasoning and mathematical proof in Grade 6 elementary school students. *Canadian Journal of Science, Mathematics, and Technology Education* 13: 70–89.

Francisco, J. M. and Maher, C. A. (2005). Conditions for promoting reasoning in problem solving: Insights from a longitudinal study. *The Journal of Mathematical Behavior* 24 (3–4): 361–372.

Francisco, J. M., and Maher, C. A. (2011). Teachers attending to students' mathematical reasoning: Lessons from an after-school research program. *Journal of Mathematics Teacher Education* 14(1): 49–66.

Goos, M. (2004). Learning mathematics in a classroom community of inquiry. *Journal for Research in Mathematics Education* 35: 258–291.

Goulding, M., Rowland, T., and Barber, P. (2002). Does it matter? Primary teacher trainees' subject knowledge in mathematics. *British Educational Research Journal* 28(5): 689–704.

Hanna G. and Jahnke H. N. (1996) Proof and Proving. In: A. J., Bishop, K. Clements, C. Keitel, J. Kilpatrick, and C. Laborde C. (eds), *International handbook of mathematics education*, 877–908. Dordrecht: Springer.

Hanna, G. (2000). Proof, explanation and exploration: An overview. *Educational Studies in Mathematics* 44 (1–2): 5–23.

Hanna, G. and de Villiers, M. (2008). ICMI study 19: Proof and proving in mathematics education. *ZDM Mathematics Education* 40: 329–336.

Hoffman, B. L., Breyfogle, M. L., and Dressler, J. A. (2009). The power of incorrect answers. *Mathematics Teaching in the Middle School* 15: 232–238.

Kazemi, E. and Stipek, D. (2009). Promoting conceptual thinking in four upper-elementary mathematics classrooms. *Journal of Education* 189 (1–2): 123–137.

Knudsen, J., Lara-Meloy, T., Stevens, H. S., and Rutstein, D. W. (2014). Advice for mathematical argumentation. *Mathematics Teaching in the Middle School* 19: 494–500.

Knuth, E. J. (2002). Secondary school mathematics teachers' conceptions of proof. *Journal for Research in Mathematics Education*, 379–405.

Kosko, K. W., Rougee, A., and Herbst, P. (2014). What actions do teachers envision when asked to facilitate mathematical argumentation in the classroom? *Mathematics Education Research Journal* 26: 459–476.

Krauss, S., Brunner, M., Kunter, M., Baumert, J., Blum, W., Neubrand, M., and Jordan, A. (2008). Pedagogical content knowledge and content knowledge of secondary mathematics teachers. *Journal of Educational Psychology* 100(3): 716–725.

Krummheuer, G. (2015). Methods for reconstructing processes of argumentation and participation in primary mathematics classroom interaction. In: *Approaches to qualitative research in mathematics education*, 51–74. Dordrecht: Springer.

Lin, P. J. and Tsai, W. H. (2012). Fifth graders mathematics proofs in classroom contexts. *Proceedings of PME*, 36.

Makar, K., Bakker, A., and Ben-Zvi, D. (2015). Scaffolding norms of argumentation-based inquiry in a primary mathematics classroom. *ZDM Mathematics Education* 47: 1107–1120.

Marks Krpan, C. (2001). *The write math*. Parsippany: Pearson Education.

Marks Krpan, C. (2011) Engaging struggling learners in group problem solving in mathematics In: C. Rolheiser (ed.), *School university partnerships: Creative connections, OISE*, 110–117. Toronto: OISE.

Marks Krpan, C. (2013). *Math expressions: Promoting problem solving and mathematical thinking through communication.* Toronto: Pearson Education.

Marks Krpan, C. (2018). *Teaching math with meaning; Cultivating self-efficacy through learning competencies.* Toronto: Pearson Education.

Marks Krpan, C. and Sahmbi, G. (2019). Arguing for access: Teachers' perspectives on the use of argumentation in elementary mathematics and its impact on student success. In: *2019 Conference of the Canadian Society for the Study of Education.*

Marks Krpan, C. and Sahmbi, G. (2020). Reasoning Represented: Teachers' perspectives on students use of mathematical representations in counter arguments in mathematics. Canadian Society for the Study of Education Conference, London, Ontario.

McCrone, S. S. (2005). The development of mathematical discussions: An investigation in a fifth-grade classroom. *Mathematical Thinking and Learning* 7: 111–133.

McCrory, R. and Stylianides, A. J. (2014). Reasoning-and-proving in mathematics textbooks for prospective elementary teachers. *International Journal of Educational Research* 64: 119–131.

Mercer, N. (2010). The effective use of talk in the classroom. *Interacció comunicativa i ensenyament de llengües*, 19–28.

Michaels, S., O'Connor, C., and Resnick, L. B. (2008). Deliberative discourse idealized and realized: Accountable talk in the classroom and in civic life. *Studies in Philosophy and Education* 27: 283–297.

National Council of Teachers of Mathematics. (2000). *Principles and standards for school mathematics.* Reston, VA: National Council of Teachers of Mathematics.

Ontario Ministry of Education (2005a). *The Ontario curriculum grades 1-8: Mathematics.* http://www.edu.gov.on.ca/eng/curriculum/elementary/math18curr.pdf

Ontario Ministry of Education. (2005b). *The Ontario curriculum grades 9 and 10: Mathematics.* http://www.edu.gov.on.ca/eng/curriculum/secondary/math910curr.pdf.

Ontario Ministry of Education. (2007). *The Ontario curriculum grades 11 and 12: Mathematics.* http://www.edu.gov.on.ca/eng/curriculum/secondary/math1112currb.pdf.

Reid, D., Knipping, C., and Crosby, M. (2011). Refutation and the logic of practices. *PNA* 6: 1–10.

Rumsey, C. W. (2012). *Advancing fourth-grade students' understanding of arithmetic properties with instruction that promotes mathematical argumentation.* Illinois State University.

Rumsey, C. (2013). A Model to Interpret Elementary-School Students' Mathematical Arguments. In *Proceedings of the 37th Conference of the International Group for the Psychology of Mathematics Education.*

Rumsey, C. and Langrall, C. W. (2016). Promoting mathematical argumentation. *Teaching Children Mathematics* 22: 412–419.

Rumsey, C., Guarino, J., Gildea, R., Cho, C. Y., and Lockhart, B. (2019). Tools to support K–2 students in mathematical argumentation. *Teaching Children Mathematics* 25: 208–217.

Schwarz, B. B., Hershkowitz, R., and Prusak, N. (2010). Argumentation and mathematics. *Educational dialogues: Understanding and promoting productive interaction* 115: 141.

Shulman, L. S. (1986). Those who understand: Knowledge growth in teaching. *Educational Researcher* 15: 4–14.

Solar, H., Ortiz, A., Deulofeu, J., and Ulloa, R. (2020). Teacher support for argumentation and the incorporation of contingencies in mathematics classrooms. *International Journal of Mathematical Education in Science and Technology.* DOI: https://doi.org/10.108 0/0020739X.2020.1733686

Solomon, A. (2009). The use of vocabulary in an eighth grade mathematics classroom: Improving usage of mathematics vocabulary in oral and written communication. *Action Research Projects* 29. https://digitalcommons.unl.edu/mathmidactionresearch/29.

Sowder, L. and Harel, G. (1998). Types of students' justifications. *The Mathematics Teacher* 91: 670–675.

Staples, M. and Newton, J. (2016). Teachers' contextualization of argumentation in the mathematics classroom. *Theory into Practice* 55: 294–301.

Stylianides, A. J. (2007a). Proof and proving in school mathematics. *Journal of Research in Mathematics Education* 38: 289–321.

Stylianides, A. J. (2007b). The notion of proof in the context of elementary school mathematics. *Educational Studies in Mathematics* 65: 1–20.

Stylianides, A. J. (2016). *Proving in the elementary mathematics classroom.* Oxford University Press.

Stylianides, G. J., Stylianides, A. J., and Shilling-Traina, L. N. (2013). Prospective teachers' challenges in teaching reasoning-and-proving. *International Journal of Science and Mathematics Education* 11: 1463–1490.

Turner, J. C. and Meyer, D. K. (2004). A classroom perspective on the principle of moderate challenge in mathematics. *The Journal of Educational Research* 97: 311–318.

Weber, K. (2014). What is a proof? A linguistic answer to an educational question. In: *Proceedings of the 17th Annual Conference on Research in Undergraduate Mathematics Education,* 1145–1149.

Whitenack, J, and Yackel, E. (2002). Making mathematical arguments in the primary grades: The importance of explaining and justifying ideas. *Teaching Children Mathematics* 8: 524–527.

Yackel, E. (2001). Explanation, justification and argumentation in mathematics classrooms. In: M. Van den Heuvel-Panhuizen (ed.), *Proceedings of the 25th conference of the international group for the psychology of mathematics education PME-25,* vol. 1. (1–9). Utrecht.

Yackel, E. (2004). Theoretical perspectives for analyzing explanation, justification and argumentation in mathematics classrooms. *Journal of the Korea Society of Mathematical Education Series D: Research in Mathematical Education* 8: 1–18.

Yackel, E. and Hanna, G. (2001). Explanation, justification and argumentation in mathematics classrooms. In: *Proceedings of the Conference of the International Group for the Psychology of Mathematics* Education (25th, Utrecht, The Netherlands).

Zolkower, B. and Shreyar, S. (2007). A teacher's mediation of a thinking-aloud discussion in a 6th grade mathematics classroom. *Educational Studies in Mathematics* 65: 177.

Chapter 4
Subverting Stereotypes: Visual Rhetoric in the #SheCanSTEM Campaigns

Deborah J. Danuser

Introduction

Research shows that young girls like STEM subjects—science, technology, engineering and math—but, as they get older, they start to feel that STEM isn't for them based on outdated stereotypes (Ad Council 2018a).

On the homepage of SheCanSTEM.com in the spring of 2019, visitors are greeted by a photograph of seven adult women positioned shoulder to shoulder, and confidently looking directly at the viewer. The caption below this "hero image" tells visitors, "#SheCanSTEM. Meet the women changing the world with Science, Technology, Engineering and Mathematics. The future will be built by women in STEM." The smiling women in the ad are the faces of the 2018 Ad Council's public service announcement (PSA) campaign, "She Can STEM," which promotes the science, technology, engineering, and mathematics (STEM) fields to tween girls (11–15 years-old). According to the campaign's webpage on AdCouncil.org, the campaign is designed to inspire "middle school girls to stay in STEM by showcasing female role models across a variety of STEM fields" (Ad Council 2018b). As such, the stars of the campaign are not professional models hired for a photoshoot, but are seven women "currently dominating the world of STEM" (She Can STEM 2019).

But why is such a campaign needed? Ryan Noonan writes in the executive summary of the U.S. Department of Commerce's report, *Women in STEM: 2017 Update* (2–17: 1): "While women continue to make gains across the broader economy, they remain underrepresented in STEM jobs and among STEM degree holders." The report goes on to list the following statistics:

D. J. Danuser (✉)
Department of Communication, University of Pittsburgh, Pittsburgh, PA, USA

© Springer Nature Switzerland AG 2020
S. A. Costa et al. (eds.), *Mathematics (Education) in the Information Age*,
Mathematics in Mind, https://doi.org/10.1007/978-3-030-59177-9_4

- Women filled 47% of all U.S. jobs in 2015 but held only 24% of STEM jobs. Likewise, women constitute slightly more than half of college-educated workers but makeup only 25% of college-educated STEM workers.
- Women with STEM degrees are less likely than their male counterparts to work in a STEM occupation; they are more likely to work in education or healthcare.
- Women with STEM jobs earned 35% more than comparable women in non-STEM jobs – even higher than the 30% STEM premium for men. As a result, the gender wage gap is smaller in STEM jobs than in non-STEM jobs. Women with STEM jobs also earned 40% more than men with non-STEM jobs (Noonan 2017: 1).

These statistics demonstrate how women are underrepresented in receiving STEM degrees and working in STEM jobs, despite the increased earning potential of STEM jobs.

This overall lack of women in STEM, combined with the fact women with STEM degrees are consigned to the educational and health fields, creates a dearth of role models for young girls interested in STEM. "As girls look around for female role models, they don't see anyone who looks like they do. If we want girls to succeed in STEM, we have to show them it's possible" (Ad Council 2018c: 2). "She Can STEM aims to challenge obsolete stereotypes and help middle school girls overcome their perceptions of what STEM isn't by surprising them with what it is" (Ad Council 2018e: 1).

Drawing upon theories of visual rhetoric and images in advertising, this research looks at how the Ad Council's "She Can STEM" campaign promotes STEM to girls. First, I contend that the campaign materials actively strive to subvert culturally-dominant stereotypes that science is a masculine endeavor by avoiding the stereotypes' most common tropes in the campaign's images. Second, I examine select "She Can STEM" campaign images via Birdsell and Groarke's (2007) modes of visual meaning. Third, I identify shortcomings in the campaign that arise from stripping its role models of all visual cues that they are scientists, as well as its exclusion of role models from the academic and government sectors.

Prevailing (Visual) Stereotypes of Scientists

When it comes to the images and ideas Americans associate with scientists (and STEM), research has shown that we hold complex, multilayered feelings. Mead and Métraux's (1957) landmark findings presented researchers with the first insights into ideas held by the public regarding science. These positive, negative, and shared images collected by Mead and Métraux about scientists have served as the foundation for almost every study that has followed. Their pilot study analyzed essays written by high school students on what they think about scientists and what kind of scientist he would or would not like to be. It is relevant to note that in the Mead and Métraux's study, female high school students were not asked what type of scientist they themselves would like to be, but rather what type of scientist their husband would probably like to be. Specifically, their research found the high school students held shared, neutral images of scientists that revolve around the scientists' appearance and physical surroundings. The positive side of scientists begins to emerge when students described some aspects of the personality, characteristics, and motivation of scientists. Good scientists are intelligent, benevolent, hardworking, and focused men working to better the world by understanding it. However, the scientists' negative side described by the students stem from the same concepts (i.e. intelligence, motivation, dedication, etc.) that are articulated in the positives, but the students associate a decidedly different value to them. These scientists are powerless, alienated, selfish and obsessive men unable to concern themselves with non-science things.

Drawing on the stereotypes Mead and Métraux articulated, Chambers (1983) developed another major contribution to the area of scientific stereotypes research—the "Draw-a-Scientist-Test (DAST)." The test consisted of having a regular classroom teacher ask elementary students to "draw a scientist" without any previous discussion or working collectively. The drawings were then analyzed and scored based upon seven previously chosen indicators of the standard image of a scientist: (1) labcoat, (2) eyeglasses, (3) facial hair, (4) symbols of research (scientific instruments and laboratory equipment), (5) symbols of knowledge (books and filing cabinets), (6) technology (the products of science), and (7) relevant captions of formulas, taxonomic classifications, etc. (Chambers 1983: 258). Chambers found students began to incorporate the elements of a stereotypical scientist starting in the second grade. As the children neared the end of elementary school, the more stereotypical their images became.

Finson et al. (1995) expanded Chambers' 7 indicators to 16 categories to create the DAST-Checklist. Finson et al's DAST-Checklist records additional common images appearing in the drawings of scientists. Categories 1–7 are identical to Chambers' DAST indicators, but the DAST-Checklist adds (8) male gender, (9) white, (10) indications of danger, (11) presence of light bulbs, (12) mythic stereotypes (Frankenstein creatures, Jekyll/Hyde figures, etc.), (13) indications of secrecy (signs saying "private, keep out, top secret," etc.), (14) scientists doing work indoors, (15) middle-aged or older scientist, and (16) open comments (dress items, neckties/necklaces, hair style/grooming, smile or frown, stoic expression, bubbling liquids, type of scientist, etc.) (Finson et al. 1995: 199).

Various versions of the DAST and DAST-Checklist have often been repeated using different subject groups ranging from elementary school students to college students and adults (Boylan et al. 1992; Finson et al. 1995; Huber and Burton 1995; Mason et al. 1991; Miele 2014; Rahm and Charbonneau 1997; Rosenthal 1993; Rubin et al. 2003; Sumrall 1995; Thomas et al. 2006). The results, even the ones drawn by scientists, consistently yield the same dominant image of a scientist learned as a child—a white male, wearing glasses and a white labcoat, who works alone in a laboratory surrounded by chemistry equipment. However, studies have shown that the older DAST participants were the more likely to draw alternate images of scientists (female, working outdoors, minority, etc.). They were also less likely to use mythic stereotypes. However, these few alternate-image drawings were usually created by minority group members, females, scientists, or by participants of science education intervention programs designed to breakdown stereotypes and make science more accessible to minorities.

One reason for the pervasiveness of the stereotypical image may be because it "reflects reality perhaps in part, but certainly not in totality" (Rahm and Charbonneau 1997: 777). Rahm and Charbonneau (1997) conducted an informal visual survey of an atmospheric research center to look for stereotypical characteristics in real scientists. They found 42% of the scientists wore glasses and 38% had facial hair/extravagant hairstyles in comparison to their DAST results of 70% and 52% for each figure respectively. However, computers and/or workstations were seen in 98% of the visual survey while they appeared in only 4% of the DAST drawings.

Whether positive, negative or neutral, the stereotypes and misconceptions that shroud science are established at an early age and linger throughout life. The development of stereotypical science images by the end of elementary school (Chambers 1983) coincides with the retreat of girls and minorities from science in secondary school (Kelly 1982). Kelly cites three main reasons girls withdraw from science, but they can be applied to ethnic minorities as well—lack of self-confidence and fear that it is too difficult, the masculine image of science, and the apparent remoteness of science from everyday concerns. Ultimately, the underlying issue of self-image is key; if a child's idea of self doesn't match his or her image (and the images provided by home, school, and the media) of a scientist, then his or her interest in science isn't proper or appropriate (Steinke 1998).

Visual Rhetoric and Advertising

Birdsell and Groarke (2007: 103) argue that visual arguments can be "understood and assessed" through Aristotle's rhetorical proofs (*ethos, pathos* and *logos*) just as traditional verbal arguments. They also list five functions that visual images can perform in a visual argument: flags, demonstrations, metaphors, symbols and archetypes. Flags are "used to attract attention to a message conveyed to some audience" (Birdsell and Groarke 2007: 104). Images act as demonstrations by conveying "information which can best be presented visually" and serve as metaphors by communicating "some claim figuratively, by portraying someone or something as some other thing" (p. 105). Symbols exhibit "strong associations that allow them to stand for something they represent," while archetypes are symbols that derive meaning from "popular narratives" (p. 105).

These five functions can be applied to visual advertisements (such as those found in print, television and online mediums) as well. For example, Henrik Dahl, as cited in *Visual persuasion: The role of images in advertising* (Messaris 1996: 5), states that a fundamental aspect of advertising is that it is normally an "*unwanted* communication." This is because consumers do not actively seek out advertisements and commercials for consumption. Therefore, a critical role of an advertisement is to flag (attract and maintain) the attention of customers. An organization's logo—a distinctive signature, motto, image, or trademark—taps into the metaphor, symbols and archetypes functions.

Once the viewers' attention is attained via the flag, Messaris argues the next step is to elicit emotions from the viewers. This can be done in a number of ways, including via Messaris's three major roles images play in advertising: (1) images as simulated reality, or iconicity; (2) images as evidence, or indexicality; and (3) images as an implied selling proposition or syntactic indeterminacy. Iconicity is critical in advertising because it allows advertisers to simulate, as well as violate, reality. Messaris (1996: xiii) argues this is important for the following reason:

> When we look at the real world that surrounds us, the sights we see do not register in our brains as neutral, value-free data. Rather, each visual feature, from the smallest nuances of people's facial expressions to the overall physical appearance of people and places, can come with a wealth of emotional associations. These associations stem from the unique experiences of each individual in addition to the common, shared influence of culture.

Whether it is simulating or violating reality, viewers of an image assign meanings, emotions and values to what they see. In this way, iconicity also can evoke *pathos* in the viewer. "Indexicality is a critical ingredient in the process of visual persuasion whenever a photographic image can serve as documentary evidence or proof of an advertisement's point" (Messaris 1996: xvi). For example, a commercial featuring a celebrity drinking a Pepsi communicates more than a written description of the video, or a drawing/animation of the celebrity enjoying the soda. Due to the inherent nature of photo/videography to "capture" reality, viewers are more likely to believe that the celebrity actually drank the soda.

The third role touches upon the syntactic indeterminacy of images. Messaris (1996: xi) describes the difference between verbal and visual syntaxes as follows:

> [A] distinctive characteristic of verbal language is the fact that it contains words and sentence structures (a prepositional syntax) that allows the user to be explicit about what kind of connection is being proposed in such statements. An equally distinctive characteristic of visual images is the fact they do not have an equivalent of this type of syntax.

While it is tempting to view a lack of specificity when it comes to visual syntax as a negative, Messaris (1996: xxii) argues that it is precisely because of its lack of specificity that visual arguments have an open-ended nature that lends itself to an "adaptability to the meaning of persuasive images."

Beasley and Danesi (2002: 12) state that two primary practices in advertising—*positioning* and *image creation*—go "about creating … messages and anchoring them firmly into social discourse. "Positioning is the placing or targeting of a product for the right people," while image creation results in "fashioning a 'personality' for the product." For example, Mountain Dew soda is positioned primarily towards men as the majority of its advertisements feature male leads and the image it has crafted appeals to competitive male teenagers and young adults interested in computer/video games and extreme sports. Beasley and Danesi (2002: 15) also state that "advertising has become entrenched into social discourse by virtue of its widespread diffusion throughout society." Advertising's ability to tap into the everchanging, ephemerality of social discourse allows it to (Beasley and Danesi 2002: 16):

- Guarantee that newness and faddishness can be reflected in the product through adaptive change in…commercials, or in the meanings embedded in its logo, package design, etc.;
- Ensure that any changes in social trends…also be reflected in ads, commercials, logos, design, etc.;
- Ensure that the product's identity keeps in step with the times by renaming it, redesigning its appearance, changing its advertising textuality, etc.;
- Guarantee that the consumer's changing needs and perceptions be built into the textuality (form and content) of [the] brand…, thus creating a dynamic interplay between advertising and changing modalities of social life, whereby one influences the other through a constant synergy.

Essentially, advertisements are so deeply integrated into our society, that popular culture takes up messages, images, themes, etc. from commercials and weaves them into itself (consumers across America saying "Whassup" after seeing Budweiser's commercials), and commercials assimilate popular culture elements like hit songs, "hip" celebrities, fashion trends, etc.

Breaking Down the "She Can STEM" Campaign

Unlike many previous Ad Council PSA campaigns, "She Can STEM" does not rely on print, radio or television components to spread its message. Instead, the campaign is focused on online media elements, such as banner ads, social media graphics, online videos, etc., that can be embedded in other websites or shared on social media. For the purposes of this chapter, I am limiting my scope of research to the seven social [media] graphics[1] produced for the campaign, which each feature a different campaign spokeswoman.

The design and composition of the social graphics are almost identical. When you look at all seven graphics, the only design differences are in the colors and which direction the woman is facing. Each one is square-shaped, which optimizes its display in some social media feeds, such as Instagram. The primary focus in the social graphics is a headshot of one of the women, which takes up approximately two-thirds of the space. In the remaining third, the letters S, T, E, M are placed in individual square boxes arranged vertically. Two of the boxes are always smaller than the others, and in the larger boxes we are able to see what the letter represents (S is for science). The two larger boxes always represent the two elements of STEM most associated with the featured woman's job. The design of letter boxes mimics

---

[1] From the "She Can STEM" campaign by the Ad Council, 2018. Retrieved April 9, 2019, from http://shecanstem.adcouncilkit.org/spread-the-word/. Copyright 2018 by the Ad Council.

that of the periodic table of elements, an iconic image found in science classrooms around the world. All but two of the women have an enlarged T in their boxes. They also are the two women who are not employed by corporate campaign partners. The M is also enlarged in only two images. The S is enlarged three times and the E four times.

Underneath the STEM boxes, the Ad Council logo appears in light gray so it can clearly brand the campaign without distracting from its core messages. In the bottom fourth of the square, there is a blue, orange or yellow box that contains an inspirational tagline from the woman in the ad to the campaign's target audience. Below are the taglines (Ad Council 2018d):

- You have the power to bring new worlds to life.
- Do what you love and you'll always be successful.
- If you can imagine it, it's possible.
- Ever wonder if there's life on other planets?
- Don't just solve the problem, write the code.
- We need girls like you in STEM.
- You are the generation that will be stepping foot on Mars.

Also written in the boxes, directly underneath the inspirational sentence, we find the featured woman's name and credentials written in a smaller font size. Nowhere in the design of the social graphics does the name of the campaign, its slogans, or references to additional information (such as a hashtag, URL address, etc.) appear.

The women were photographed in front of a solid, white background. They are making direct eye contact with the camera (and therefore the viewers) and are smiling. The camera's angle is perpendicular to the subjects and not shot using high or low angles. The images are cropped to include the head, shoulders, torso and occasionally, the waist. Many of the women are posed similarly in their headshots. Some have their arms crossed in front of their chests. One is posed with both her hands on her hips, while another has only one hand on her hip. The arms of another woman are at her sides, but her hands are laying on top of one another in a manner that suggests they are resting on something just out of frame. The women are wearing either casual attire (a long-sleeved t-shirt, a denim shirt, or a motorcycle jacket with a V-neck t-shirt) or business-casual attire (dress blouses with and without jackets). They are all wearing subtle make-up, and none of them have their hair pulled back. One woman's hair color is noticeable as it is dyed a soft pink while the rest of the campaign spokeswomen's hair colors appear natural.

To better analyze these visuals, I modified the DAST-Checklist created by Finson et al. to create the Visual Stereotypes of Scientists Checklist (VSSC). The VSSC streamlines some aspects of the DAST-Checklist (such as putting relevant captions of formulas and taxonomic classifications within the symbols of knowledge category), while moving others out of the "open comments" section into their own category (such as stoic expression). These changes make it easier to score images of scientists based on how many of the following 15 visual stereotypes occur:

- Visual presents as male
- Visual presents as white
- Stoic expression present
- Facial hair is present (if male) or long hair is pulled back (if female)
- Appears to be a middle-age or older adult
- Eyeglasses are worn
- Specialized attire or clothing (labcoats, goggles, clean suits, protective gloves, etc.) is worn
- Appearance is disheveled (messy hair, clothing askew, etc.)
- Mythic stereotypes (Frankenstein creatures, Jekyll/Hyde figures, Albert Einstein, etc.) are present
- Set indoors
- Symbols of research (scientific instruments and laboratory equipment) appear.
- Symbols of knowledge (books, filing cabinets, lightbulbs, relevant captions of formulas, taxonomic classifications, etc.) appear
- Technology (the products of science and engineering, such as computers) is present
- Indications of danger are visable
- Indications of secrecy appear

For example, there is one telling image of a scientist taken from a stock photography website[2], which scores an 11 on the VSSC—male, white, facial hair, middle-age/senior adult, eyeglasses, specialized attire/clothing (labcoat, goggles, protective gloves), disheveled appearance (specifically his unkempt hair), working indoors, symbols of research (beakers, mixing chemicals, etc.), symbols of knowledge (formulas on the chalkboard, pen and paper for taking notes, etc.), and indication of danger (gas mask for filtering out dangerous fumes and protective gloves for corrosive chemicals). Even a less cartoonish stock photo[3] scores a 9 on the VSSC.

Implications

As previously mentioned, the goal of the "She Can STEM" campaign is to keep girls interested in STEM by providing girls with female role models that do not reflect outdated stereotypes. The predominant stereotypical image of a scientist is an older, white male with facial hair and wild, unkempt hair that is wearing glasses and a labcoat. He stoically works indoors surrounded by scientific instruments (such

---

[2] "Mad scientist conducts chemistry experiment in his lab" by J. McRight, 2019. Retrieved from https://www.shutterstock.com/image-photo/mad-scientist-conducts-chemistry-experiment-his-113472703https://.

[3] "Scientist using microscope in a modern laboratory" by caracterdesign, 2019. Retrieved from https://www.gettyimages.com/detail/photo/scientist-using-microscope-royalty-free-image/181892523.

Table 4.1 Visual stereotypes of scientists checklist (VSSC) scores for 9 figures

Stereotype	1	2	3	4	5	6	7	8	9
Male								1	1
White	1	1	1	1		1		1	1
Stoic expression									1
Facial hair								1	1
Middle-age, older adult								1	1
Eyeglasses								1	1
Specialized attire/clothing								1	1
Disheveled appearance								1	
Mythic stereotypes									
Working indoors	1	1	1	1	1	1	1	1	1
Symbols of research								1	1
Symbols of knowledge								1	
Technology									
Indication of danger								1	
Indication of secrecy									
SCORE	2	2	2	2	1	2	1	11	9

as a microscope) and chemical laboratory equipment. As such, the advertising professionals who worked on the "She Can STEM" campaign actively subvert this stereotype by eliminating as many of these characteristics as possible from its campaign materials.

In stark contrast to the culturally-dominant stereotype, the role models in the "She Can STEM" campaign are all women who look to be in their 30s and are wearing casual or business-casual attire. Their relative youth and lack of specialized attire help make the women more accessible and relatable to the campaign's target audience of tween girls. The plain white background and lack of props in the photos also removes stereotypical symbols from the campaign materials. This encourages girls to imagine what the women's work environments looks like and to place the women in an environment of their choosing.

The campaign images are so a-stereotypical that the five of the seven social graphics examined scored a 2 on the VSSC while the remaining two scored a 1 (see Table 4.1). All seven graphics earned a point for "working indoors" as the plain white background implies an interior environment. The subsequent points earned by each graphic depended on if the woman was white (5 of 7 were).

Another function of the white background in the social graphics is that it helps attract the viewer's eye. Or as Birdsell and Groarke (2007) would reason, the white space functions as a *flag*. Another function of visual argument performed by the images in the graphics is *demonstration*. The photographic images of real women who work in prestigious STEM jobs demonstrate that STEM isn't just the domain of men; women can and do succeed in STEM careers. Similarly, these images of women also evoke Messaris's concepts of *iconicity* and *indexicality* as they simulate reality and offer proof of success in STEM.

An aspect of the campaign that is notable is its purposeful avoidance of *symbols* and *archetypes*. The women are not holding, nor are surrounded by, any props or workplace imagery symbolically associated with STEM. The lack of *symbols*, combined with the gender and youth of the women, effectively avoids the *archetypes* of science. In fact, the images score so low on the VSSC that if it was not for the accompanying text/copy in the social graphics, viewers would not know that the women work in STEM. It could be said that the STEM boxes in the campaign materials act as symbols since their design reminds viewers of the periodical table and present the STEM acronym. However, I argue that the STEM boxes alone do not definitively link the women in the pictures to STEM jobs.

The accompanying copy tells us that these women work for IBM, Google, Microsoft, Verizon, Boeing, GE and the Adler Planetarium. All of these employers are multinational corporations except the Adler Planetarium, which is a non-profit organization. Five of the seven employers are brand partners of the campaign (IMB, Google, Microsoft, Verizon and GE). Representative of the women who work in academic and government STEM sectors are conspicuously absence. This is surprising as approximately 30% of the STEM workforce is employed by the education and government sectors (National Science Board 2018).

Since advertisements are unwanted communications, flagging (and retaining) the attention of the target audience must be immediate. As previously stated, I believe the large amounts of white space in the campaign materials act as a flag, but the lack of science symbols does not give the tweens interested in STEM a reason to linger on the campaign's social graphic. The STEM boxes may help increase the target audience's attention, but ultimately, the brain must decide in a fraction of a second if the combination of the white space, the image of a woman, and the nearby STEM boxes is enough to make the viewer stop and read the ad's copy. If it is not compelling enough to the viewer, the brain will filter out the ad as background noise like it does to most ads.

The campaign strips away all of the visual cues that these women are scientists and instead relies on text/body copy to convey that information. In actively striving to subvert culturally-dominant visual stereotypes of scientists, the campaign is actually ignoring a whole sector of women who work in STEM—the women that do wear labcoats, work with microscopes, monitor chemical reactions, etc. The campaign does not show women doing "stereotypical" science that may be associated with the government and academic sectors. Instead, it showcases tech-based jobs found at its (corporate) brand partners. In short, we lose the immediate recognition that she is a scientist because the campaign overcompensates to ensure that image is not stereotypical. This could have been avoided by giving the women simple props to interact with during the photoshoot. For example, the woman astronomer, could be leaning on a backyard-sized telescope. Another woman could be holding a video game controller since she heads the Halo Game Studio at Microsoft. Instead of standing in front of a white background, a third woman could be standing in front of computer code, as she tells girls, "Don't just solve the problem, write the code."

Another shortcoming of the campaign's social graphics is that they do not include any references to the campaign's slogan ("She can STEM, so can you"), its primary hashtag (#SheCanSTEM), website URL (SheCanSTEM.com) or its Instagram account handle (@SheCanSTEM). I suspect this is because text appearing in a picture cannot be hyperlinked, so even if the image included #SheCanSTEM, users could not click on it to follow a link. However, the lack of the campaign information as problematic as "She Can STEM" is designed to be primarily an online campaign. Not including references to where the campaign lives online places the burden of sharing that information on the social media user. If the user doesn't include #SheCanSTEM, @SheCanSTEM, or SheCanSTEM.com, then viewers of the post do not know it is a part of a large campaign.

Conclusion

Advertisements play critical role in our culture as they attempt to persuade us buy a particular product or take a particular action. The visual arguments created by the images in advertisements serve as a flag, demonstration, symbol, archetype and/or metaphor for the viewers (Birdsell and Groarke 2007) and can elicit emotion by evoking iconicity, indexicality, and syntactic indeterminacy (Messaris 1996). Successful advertising campaign both become part of our social discourse, as well as influence it (Beasley and Danesi 2002).

The Ad Council has a rich history of producing memorable public service campaigns, including Smoky the Bear, the Crash-Test Dummies (who remind us to "Don't be a dummy, buckle your seat belt,"), and more. The Ad Council's 2018 "She Can STEM" campaign was created to address the fact that girls start to lose interest in STEM in middle school due in part to a cultural belief that STEM is a masculine endeavor. The campaign sought to counter this narrative by showcasing real women who work in STEM as role models for tween girls.

The campaign's materials, specifically its social graphics, successfully subvert the majority of scientist stereotypes in order to present tween girls with contemporary role models. However, it subverts to the point there are no visual cues that the featured women are scientists and the audience must rely on accompanying copy to understand their connection to STEM. The social graphics also fail to include relevant online information (i.e., URLs, hashtags, handles), which is troubling as the campaign is designed to be a social media campaign. Finally, the campaign ignores STEM careers in the academic and government sectors. Instead, it relies almost exclusively on corporate campaign partners.

References

Ad Council. (2018a). Empowering Girls in STEM. Ad Council. Retrieved April 9, 2019, from https://www.adcouncil.org/Our-Campaigns/Education/Empowering-Girls-in-STEM

Ad Council. (2018b). Girls in STEM: Campaign Overview. Retrieved April 9, 2019, from http://shecanstem.adcouncilkit.org/wp-content/uploads/sites/66/2018/09/STEM_Overview_9.18.18.pdf

Ad Council. (2018c). She Can STEM Campaign Talking Points. Retrieved April 9, 2019, from http://shecanstem.adcouncilkit.org/wp-content/uploads/sites/66/2018/09/She-Can-STEM-Talking-Points.pdf

Ad Council. (2018d). Spread the Word. Retrieved April 9, 2019, from http://shecanstem.adcouncilkit.org/spread-the-word/

Ad Council. (2018e). Women Constitute 50 Percent of US College-Educated Workforce, but Only 25 Percent of STEM Workforce; GE, Google, IBM, Microsoft and Verizon Join Forces to Empower Young Girls to Pursue Science, Technology, Engineering and Math (STEM) in New Ad Council Campaign. Retrieved April 9, 2019, from http://shecanstem.adcouncilkit.org/wp-content/uploads/sites/66/2018/09/She-Can-STEM-Press-Release-Addendum.pdf

Beasley, R. and Danesi, M. (2002). *Persuasive signs: the semiotics of advertising.* Berlin: Mouton de Gruyter.

Birdsell, D. S., and Groarke, L. (2007). Outlines of a theory of visual argument. *Argumentation and Advocacy* 43: 103–113.

Boylan, C. R. Hill, D. M., Wallace, A. R., and Wheeler, A. E. (1992). Beyond stereotypes. *Science Education* 76: 465–476.

Chambers, D. W. (1983). Stereotypic images of the scientist: The draw-a-scientist test. *Science Education* 67: 255–265.

Finson, K. D., Beaver, J. B., and Cramond, B. L. (1995). Development and field test of a checklist for the draw-a-scientist test. *School Science and Mathematics* 95: 195–205.

Huber, R. A. and Burton, G. M. (1995). What do students think scientists look like? *School Science and Mathematics* 95: 371–376.

Kelly, A. (1982). Why girls don't do science. *New Scientist* 94 (1306): 497–500.

Mason, C. L., Kahle, J. B., and Gardner, A. L. (1991). Draw-a-scientist test: Future implications. *School Science and Mathematics* 91: 193–198.

Mead, M., and Métraux, R. (1957). Image of the scientist among high-school students. *Science* 126 (3270): 384–390.

Messaris, P. (1996). *Visual persuasion: the role of images in advertising.* London: Sage Publications.

Miele, E. (2014). Using the draw-a-scientist test for inquiry and evaluation. *Journal of College Science Teaching* 43: 36–40.

National Science Board. (2018). *Science and engineering indicators 2018.* (NSB-2018-1). National Science Foundation.

Noonan, R. (2017). *Women in STEM: 2017 update.* U.S. Department of Commerce, Office of the Chief Economist, Economics and Statistics Administration.

Rahm, J., and Charbonneau, P. (1997). Probing stereotypes through students' drawings of scientists. *American Journal of Physics* 65: 774–778.

Rosenthal, D. B. (1993). Images of scientists: A comparison of biology and liberal studies majors. *School Science and Mathematics* 93: 212–216.

Rubin, E., Bar, V., and Cohen, A. (2003). The images of scientists and science among Hebrew- and Arabic-speaking pre-service teachers in Israel. *International Journal of Science Education* 25: 821–846.

Steinke, J. (1998). Connecting theory and practice: Women scientist role models in television programming. *Journal of Broadcasting and Electronic Media* 42: 142–151.

Sumrall, W. J. (1995). Reasons for the Perceived Images of Scientists by Race and Gender of Students in Grades 1–7. *School Science and Mathematics* 95: 83–90.

Thomas, M. D., Henley, T. B., and Snell, C. M. (2006). The draw a scientist test: A different population and a somewhat different story. *College Student Journal* 40: 140.

Chapter 5
Graphical Literacy, Graphicacy, and STEM Subjects

Stacy A. Costa

Introduction

Ever since Descartes invented analytic geometry, graphs have become integral to mathematics and science. Graphic literacy is even more important today, as students are exposed continuously to graphic artifacts; as a result, discerning between all kinds of graphs and scientific ones is a critical skill, given that the latter is intrinsic to STEM subjects. Therefore, equipping them with graphical literacy is a first step in providing students with knowledge methods to leverage, demonstrate, and apply their creative endeavours and to approach and solve hard problems. Within the (STEM) realm, students should, in a phrase, be visualizing certain facts or phenomena in terms of how they are represented graphically. While there are opportunities, even if few, throughout STEM classrooms to develop graphical literacy, typically, it is assumed to emerge spontaneously. Nevertheless, this is not necessarily the case. By practising and reinforcing graphical literacy concretely, students are given the opportunities required to extract structural and unstructured information correctly and meaningfully from graphical representations.

This chapter will consider why these opportunities are essential to promote proficient learners in STEM subjects. Furthermore, it examines problematic and common misconceptions about graphs, as well as inaccuracies related to creating them. Students should be able to utilize graphs in varied ways to understand the data, to be able to reason, communicate, and to make predictions.

S. A. Costa (✉)
Department of Curriculum, Teaching & Learning / Collaboration in Engineering Education,
Ontario Institute for Studies in Education - University of Toronto, Toronto, ON, Canada
e-mail: stacy.costa@mail.utoronto.ca

© Springer Nature Switzerland AG 2020
S. A. Costa et al. (eds.), *Mathematics (Education) in the Information Age*,
Mathematics in Mind, https://doi.org/10.1007/978-3-030-59177-9_5

Graphs in Education

Students need to be able to read graphs not only as part of mathematical learning but also to identify trends in their subject areas. Larkin and Simon (1987: 98) suggest that graphs and diagrams are superior to verbal descriptions since they "support a large number of perceptual inferences, group information through location and avoid textual searching." While graphs are essential, they are not always so simply presented. Most diagrammatic representations consist of standardized icons, which are designed to aid general understanding visually. However, less obvious, non-iconic, information may be hidden in the visual form, and this would require interpretation and inference in terms of higher-order thinking, allowing the student to observe trends or patterns that need intrepretation. If students can only read a graph at a surface level by understanding, say, the role of the x- and y-axes, but are not provided with additional insights related to how graphs provide information by visual techniques, they may miss the importance of a graph in terms of its representational power. Diagrams in general, allow us to make all kinds of generalizations. Moreover, it is vital to understand, that when reading or creating a graph, what is represented and what we want to represent may not exactly match. The clarity in representation is always an objective to be achieved in any pedagogical context. The ability to see below the graphic text to decode what they may contain in terms of assumptions is too.

Graphs can be problematic, and misconceptions can easily arise because students assume that they contain all the information required objectively. They need to see implications in the graphic structure and thus assess what kinds of empirically-based and argumentative claims are involved. A graph is, at one level, a tool used to communicate relevant scientific-mathematical ideas visually. In principle, it should consist of reliable information synthesized graphically so that decisions about something can be made realistically and empirically. Graphs can, of course, be mathematically sophisticated, but even so, for most practical purposes, they should be clear and effective in their representational formats. However, this is not always the case. Distinguishing between representation and misrepresentation thus requires the development of graphical literacy, which is not only based solely on understanding the objective of the representation, but also on what inherent knowledge beliefs are implicit in the graphic design. If we expect students to be multi-literate citizens today, they certainly need to make sense of graphs, given their ubiquity, decoding any potential hidden beliefs, and how these are shaped by specific graphics, visuals, numerical, and spatial dimensions.

Graphical Literacy

Graphical literacy can be defined simply as the ability to interpret and construct graphs. Fry (1981) located the development of such literacy within the field of mathematics, seeing a graph as a complex mathematical text, consisting of language, numbers, symbols, visual shapes, etc. which cohere into a holistic meaning system. However, today, graphs are found beyond this domain, constituting a visual language that is everywhere. They are psychologically powerful because they have the ability to represent data in a holistic way that induces specific kinds of interpretations. Graphs are now part of a broader human-computer interactive mode of processing large quantities of data into a visual narrative structure.

The importance of graphical literacy cannot be overstated; and it cannot be assumed tha a graph's meaning is self-evident. Graphical literacy cannot be linked simply to vision; like any language, graphs bear ideological and knowledge-specific subtexts. Therefore, graphical literacy, like verbal language literacy, is a skill that needs to be developed in the classroom, inhering in the ability to unpack the complex information inherent in graphical representations.

As students now have access to technological devices and all kinds of information at their fingertips, they are being constantly exposed to digital content ranging from text, video, visual guides, and audio. Faced with this technological melange on a daily basis, it is little wonder that students have become accustomed unconsciously to the visual mode of information conveyance. They can find an answer in seconds and access millions of texts to further gain relevant information. To avoid the crystallization of habituated thinking in this domain, students need to be involved in developing higher-order thinking skills when processing information through appropriate pedagogy. Graphical literacy in particular should be part of this pedagogy, given that graphics which are easily accessible online may be overlooked for what they entail.

Graphical literacy is especially critical in the domain of STEM subjects and how they encode knowledge. Students need to enhance their graphical literacy in order to filter and decipher what information is relevant and verifiable, before coming to an informed decision as to what a graph implies, especially since graphics can be created to be visually pleasing, and also designed to convey a false picture of specific information. Additionally, students need to acquire the ability to create graphs to represent scientific information clearly and meaningfully. Jonassen and Carr (2000) discovered that students found it challenging to represent their understanding in multiple graphic ways. So, while students may be engaged with information at a basic level, they typically cannot represent it at higher abstract levels of graphicality. By instructing students on how to translate concepts into visual-multimodal texts, they enhance their awareness of the concepts involved in the representational translation. Graphical literacy is thus a central skill to be acquired by STEM learners (Gebre 2018). It requires planning, a conscious understanding of the role of design layout, method of colour use, space, annotation, and so on.

Graphicacy

The term *graphicacy* was introduced by Balchin and Coleman (1965) as the ability to communicate spatial concepts that cannot be communicated by words or numbers alone. Wilmot (1999) defined graphicacy as the understanding of spatial concepts in terms of graphic symbolism. Fingeret (2012) argued that such competence involves the ability to shift from visual symbols alone to multimodal representations that include a plethora of graphical techniques. These range from infographics, charts, diagrams, maps, illustrations, and timelines to various subject-specific visual symbols. Teaching graphicacy today involves how to make sense of these techniques and how to extract relevant meanings from them, especially in STEM subjects which rely extensively on graphicacy as part of their modus operandi.

In STEM subjects, graphs are not optional devices, as they may be in the humanities; they are intrinsic, and students must learn how to utilize them as cognitive tools for both representing information and fleshing any hidden beliefs that are non-scientific. Needless to say, graphicacy has always been a pedagogical instrument in these fields, but it has largely been assumed that students can easily read graphs. However, this cannot be taken for granted, especially given the increasingly visual nature of communication in the current information age, and the vast amount of data and trends that are communicated visually through automated systems, including apps and other software. From this visual morass, students need to know how to sift from it what is relevant from what is not. They also need to understand the mathematical notions behind the softwares, which can be applied diversely for various purposes. Graphicacy requires, above all else, the ability to envision relationships among variables in graphical structure, much like syntax in language. In STEM subjects, this includes how variables relate to each other along graphical axes or dimensions.

Milner-Bolotin and Nashon (2012) found that biology students enhanced their knowledge after developing visual-graphic literacy, suggesting that this type of competence should be the basis for developing advanced reasoning skills, arguing that these actually fit in with a constructivist frame of knowledge acquisition (Vygotsky 1978; Piaget 1926). Visual literacy and critical thinking in this framework are intertwined and complementary to linguistic literacy. All types of literacy involve the ability to extract relevant meaning from information, as well as the awareness of how we construct texts. Some now claim that visual literacy is part of general "discursive fluency" (Airey and Linder 2009; Offerdahl et al. 2017), which is critical to how we communicate through various modalities of representation. While students may have fluency in one modality or other, they may not be able to integrate the separate modalities into an overall discursive-cognitive multimodal system. Moreover, this means going beyond traditional assumptions of what students bring to the classroom.

Soloways and Norris (2017) found that within the elementary classroom, students spent 90% of their time with text-based materials and only 10% with imaged based materials. However, outside of the classroom, the opposite was true. The researchers used the term "picting" in reference to how students today grasp and communicate information digitally, which stands in contrast to "writing." This has obvious pedagogical implications for tapping into the student's picting competence and transforming it into a more traditional writing-reflexive form of cognition. This can be done with cooperation between STEM and the humanities, such as the use of artistic and technical drawings in tandem, whereby students can begin to translate their picting competence into a broader graphicacy-literacy competence.

Interpretation

Graphicacy is an intrinsic part of relaying and representing information in all the sciences. As mentioned, though, it cannot be assumed that it is more comprehensible than concepts expressed in verbal language. Galesic and Garcia-Retamero (2011: 451) found that graphic comprehension "is not entirely intuitive but requires a certain level of meta-knowledge about graphs acquired through formal education." Pinker (1990) used the term "graph schema" in a similar fashion to describe the meta-knowledge of a graph. Whatever term we use, it is obvious that graphicacy cannot be assumed in students—a common pedagogical assumption. Graphicacy is intrinsic to complex mathematical or scientific representations (Rosenblatt 2004; Sierschynski et al. 2014). Bolotin (2015) argues that while many students can acquire visual literacy skills in courses such as fine art, they do not necessarily transfer them to the STEM field, and vice versa. Yet, without it, research is showing that divergent thinking and problem-solving skills required by STEM subjects are hampered (Tytler 2016).

Shah and Hoeffner (2002) found that systematic errors emerge when visual texts are not explicitly created to represent specific information. Students rarely describe the true intentions of the graphs accurately. For example, students might misinterpret a graph representing the speed of a race-car to mean the race-car's position on a track (Janvier 1981). This error is particularly common in contexts for which iconic interpretation is implied, especially when the graph is meant to represent change (such as growth or speed) in terms of some dimension (such as height, growth, location, speed). Shah and Hoeffner (2002) found that issues such as distance between graph lines, devices such as bar orientation in a graph, and quantified variables might be interpreted as literal "pictures" of the situation rather than abstractions.

Moreover, like linguistic styles, there are graphic styles that require particular kinds of interpretive skills. For example, a pie chart would typically be used to compare parts of a whole and not the difference between them. So, if students were looking at the separate entities, without some knowledge of how they are interrelated, errors in interpretation might emerge. A bar graph would be a better option for showing the differences between elements. Another problematic area is how graphs compress or scale information, which requires a specific kind of discernment. For example, showing data points on a graph of only a few months out of 1 year may imply an upward trend. However, a much broader data range might reveal a different picture. So, if scaling is too expanded or too compressed, the representation may lead to skewed interpretations. Axis style is also operative in producing misinterpretations. For example, if we start the vertical axis at 100, but the data points are 100, 150, and 120, the difference appears to be dramatic and significant, even though this may not be the case. By starting a vertical axis at zero, a more accurate depiction of the data is possible. Axis style is the simplest way to misrepresent the data, but it is also easily remedied.

On the positive side, Galesic and Garcia-Retamero (2011) found that graphical devices can assist students with low numeracy. Nevertheless, while their findings assess understanding of graphical formats, not all their subjects understood the visual displays, emphasizing again that graphical literacy requires some meta-knowledge. Overall, Costa (2017) found that students benefit from incorporating graphical and visualization as part of their mathematical learning, and can use them to explain their understanding of mathematical materials better. As Garcia-Retamero and Cokely (2013: 392) also discovered, unless graphical aids are "transparent, with well-defined elements, they only can represent relevant information marking part-to-whole relationships allows for improved comprehension." As Shah and Hoeffer (2002) argue, this implies that graphs can describe familiar relationships or unfamiliar relationships in concrete ways.

Conclusion

Piaget's (1926) notion of cognitive disequilibrium implies that previous knowledge schemas need to be updated continuously as the mind seeks to gain an equilibrium. Students who are engaging with graphical literacy are adopting new ways to update their schemas of understanding, leading to an equilibrium of knowledge. As Scaife and Rogers (1996) point out, this means that a more integrated teaching approach is required—one that links external and internal representations with prior knowledge; as well as creating conditions which facilitate making connections between the two. This has been called transliteracy (Liu 2012), and is defined as the ability to negotiate the varied and fragmented informational world of the Internet, extending traditional conceptions of literacy while enabling students to work more consistently and effectively with information. As Buuren et al. (2015) suggest, this implies providing students with opportunities to "express their own conceptual understanding and develop a more qualitative conceptual reasoning and analogical reasoning through the visual representations within graphical modes."

Future pedagogical study of graphical literacy should examine the methodology of how internal and external forms of graphicacy can be amalgamated and shown to have differential cognitive uses. Students may be bombarded with false information through automated and curated technologies and believe it to be true. Classroom pedagogy must therefore be able to get students to sift the misrepresentations from the graphical forms and this requires concrete instruction in graphicacy.

References

Airey, J and Linder, C. (2009). A disciplinary discourse perspective on university science learning: Achieving fluency in a critical constellation of modes. *Journal of Research in Science Teaching* 46: 27–49.

Balchin, W. G. V. and Coleman, A.M. (1965). Graphicacy should be the fourth ace in the pack. *The Times Educational Supplement*, S November.

Buuren, O., Heck, A., and Ellermeijer, T. (2015). Understanding of relation structures of graphical models by lower secondary students. *Research in Science Education* 46: 633-664.

Costa, S. (2017). *Math discourse in a Grade 2 Knowledge Building classroom.* Master's thesis, University of Toronto.

Fingeret, L. (2012). Graphics in children's informational texts: A content analysis. *ProQuest Dissertations and Theses Global* (1039317370).

Fry, E. (1981). Graphical literacy. *Journal of Reading* 24: 383-390.

Galesic, M. and Garcia-Retamero, R. (2011). Graph literacy: A cross-cultural comparison. *Medical Decision Making* 31: 444-457.

Garcia-Retamero, R. and Cokely, E. T. (2013). Communicating health risks with visual aids. *Current Directions in Psychological Science* 22: 392-399.

Gebre, E. (2018). Learning with multiple representations: Infographics as cognitive Tools for authentic learning in science literacy. *Canadian Journal of Learning and Technology* 44. https://thejournal.com/Articles/2017/05/08/Picting-Not-Writing.aspx?Page=4

Janvier, C. (1981). Use of situations in mathematics education. *Educational Studies in Mathematics* 12: 113-122.

Jonassen, D. and Carr, C. S. (2000). Mindtools: Affording multiple knowledge representations for learning. In: S. Lajoie (ed.), *Computers as cognitive tools: No more walls*, vol. 2, 165-196. Mahwah: Lawrence Erlbaum Associates.

Larkin, J. H. and Simon, H. A. (1987). Why a diagram is sometimes worth ten-thousand words. *Cognitive Science* 11: 65-99.

Liu, A. (2012). This is not a book: Transliteracies and long forms of digital attention. Paper presented at the Translittératies Conference, 2012. www.stef.ens-cachan.fr/manifs/translit/Alan_Liu_this-isnot-a-book-slides_2012_11_07.pdf._

Milner-Bolotin, M. (2015). Visual literacy skills in science, technology, engineering, and mathematics education. r In: The *SAGE Encyclopedia of Educational Technology*. myaccess.library.utoronto.ca/10.4135.

Milner-Bolotin, M. and Nashon, S. M. (2012) The essence of student visual-spatial literacy and higher-order thinking skills in undergraduate biology. *Protoplasma* 249 (Suppl 1) S25-S30.

Offerdahl, E. G., Arneson, H. B., and Byren, N. (2017). Lighten the load: Scaffolding visual literacy in biochemistry and molecular biology. *Life Sciences Education* 16: 1-11.

Piaget, J. (1926) *The language and thought of the child.* New York: Harcourt, Brace.

Pinker, S. (1990). A theory of graph comprehension. In: R. Freedle (ed.), *Artificial intelligence and the future of testing*, 73-176. Mahwah: Lawrence Erlbaum Associates.

Rosenblatt, L. M. (2004). The transactional theory of reading and writing. In: R. B. Ruddell and N. J. Unrau (eds.), *Theoretical models and processes of reading, 1363-1398.* Newark: International Reading Association.

Scaife, M. and Rogers, Y. (1996) External cognition: How do graphical representations work? *International Journal of Human-Computer Studies* 45: 185-213.

Shah, P. and Hoeffner, J. (2002). Review of graph comprehension research: Implication for instruction. *Educational Psychology Review* 14: 47-69.

Sierschynski, J., Louie, B., and Pughe, B. (2014). Complexity in picture books. *Reading Teacher* 68: 287-295

Soloway, E. and Norris C. (2017). Picting, not writing, is the literacy of today's youth. *The Journal.* https://thejournal.com/articles/2017/05/08/picting-not-writing.aspx.

Tytler, R. (2016) Challenges for mathematics within an interdisciplinary STEM education. Paper presented at the *13th International Congress on Mathematical Education*. Hamburg, Germany.

Vygotsky, L. (1978) *Mind in society: The development of higher psychological processes*. Cambridge: Harvard University Press.

Wilmot, D. (1999). Graphicacy as a form of communication. *South African Geographical Journal* 81: 91-92.

Chapter 6
Mathematics, Statistics, and Sports

Frank Nuessel

Introduction

In *Learning and teaching mathematics in the global village,* Danesi (2016: vii) points out that:

> It cannot be denied that technology today is reshaping the world, including the academy. It has also taken the academy into the world. Math is now a common theme in popular forms of entertainment (in movies, in television programs, and so on) and this incorporation into the popular imagination [...] can be turned to the advantage of classroom pedagogy. The extension of the math classroom into the world of pop culture is another example of how the wall-less classroom can unfold.

A sizeable body of literature documents Danesi's assertion (Nuessel 2012). Many of these studies verify the intermingling of popular culture and the academy as an enticing resource for the teaching of mathematics and statistics. The bibliographic references section of several representative studies on this significant matter contains a sizeable list of this type of research.

The first part of this chapter defines and discusses several basic terms related to the topic of teaching mathematics and statistics. These include the following: mathematics, mathematics pedagogy, statistics, statistics pedagogy, mathematics and statistics, mathematics anxiety and statistics anxiety, information, pop culture and its relationship to mathematics and statistics, games, sports, sports wagering, and most popular sports in North America with a brief historical overview of American football, basketball, and major league ball. The second part will discuss how sports information or data can provide the basis for mathematical and statistical problem-solving from the realm of popular culture to create Danesi's (2016: 82–83, 137) "wall-less" academy through the use of the vast amount of information from aspects of everyday life—an approach that will have great appeal to students. This chapter

F. Nuessel (✉)
University of Louisville, Louisville, KY, USA
e-mail: frank.nuessel@louisville.edu

© Springer Nature Switzerland AG 2020 73
S. A. Costa et al. (eds.), *Mathematics (Education) in the Information Age,*
Mathematics in Mind, https://doi.org/10.1007/978-3-030-59177-9_6

will provide selected examples of mathematical and statistical pedagogical problems from American football, baseball, and basketball.

Mathematics

What is mathematics? Berggren et al. (2020) define it as:

> The science of structure, order, and relation that has evolved from elemental practices of counting, measuring, and describing the shape of objects. It deals with logical reasoning, quantitative calculation, and its development has involved an increasing degree of idealization and abstraction of its subject matter. Since the 17th century, mathematics has been an indispensable adjunct to the physical sciences and technology, and in more recent times it has assumed a similar role in the quantitative aspects of the life sciences.

Devlin (2000: 5) describes mathematics simply as *"the science of patterns"* (emphasis in original). He goes on to say that (2000: 8):

> The patterns studied by the mathematician can be either real or imagined, visual, or mental, static or dynamic, qualitative or quantitative, utilitarian or recreational. They arise from the world around us, from the depths of space and time, and from the workings of the human mind. Different kinds of patterns give us different branches of mathematics. For example, number theory studies (and arithmetic uses) the patterns of number and counting; geometry studies the patterns of shape; calculus allows us to handle patterns of motion; logic studies the pattern of reasoning; probability theory deals with patterns of chance; topology studies patterns of closeness and position.

Mathematical Pedagogy

What is mathematical pedagogy? Danesi (2016: 1) provides a historical and informative account of math education from its Greek origins to today, pointing out that *Elements* by Euclid (mid-fourth century BCE to mid-third century BCE) was the first theoretical treatise on mathematics that served as a textbook. In that era, of course, the word "textbook" meant a hand copied document on parchment that required a considerable amount of time and effort by the scribes who reproduced these texts. With the advent of Johannes Gutenberg's printing press, textbooks became available in multiple copies, and, ultimately, facilitated mass education. This world order is what Marshall McLuhan called the Gutenberg Galaxy (McLuhan 1962). Danesi (2016: 52–57) subsequently introduces the term "Digital Galaxy" to refer to electronic media that provides open access to information via the World Wide Web (WWW) developed by Tim Berners-Lee in the early 1990s. This electronic-digital tool has resulted in a resource that provides immense amounts of data and information that can be updated and corrected regularly. In essence, it is an information search tool that permits interactive research communication. The World Wide Web is essentially a massive source of information for the public—one that must be used with care and discrimination in order to avoid false or inaccurate information. This digital tool provides teachers and students with vast data bases that facilitate easy access to a wide array of information, which can serve as stimulating resources that can engage students of mathematics and statistics with tantalizing applications of theory to practice. This virtual space is where the World Wide Web and the classroom converge to enthuse students and teachers alike in a mutually engaging arena for acquiring the knowledge of necessary to understand mathematics.

Statistics

What is statistics? Lexico (2020) defines statistics as "the practice or science of collecting and analyzing numerical data in large quantities, especially for the purpose of inferring proportions in a whole from those in a representative sample." Williams, Anderson, and Sweeney (2020) refer to statistics as "the science of collecting, analyzing, presenting, and interpreting data." The purpose of statistics is to provide meaningful information about specific matters in a particular field. Statistics includes several subdomains, namely, probability and stochastic processes, or random variables.

Statistical Pedagogy

What is statistical pedagogy? It is a set of procedures that instructors should introduce in their statistics courses to facilitate the organized teaching and learning of this discipline. By following these simple methods and techniques, students will acquire the basic elements of statistics through individual and collaborative approaches. Cobb's (1992: 15–18) detailed overview of the teaching of statistics allows for several observations and recommendations, which continue to be true, and are summarized here.

Recommendation I: Emphasize statistical thinking
The need for data.
The importance of data production
The omnipresence of variability
The quantification and explanation of variability

Recommendation II: More data and concepts: Less theory, fewer recipes

Recommendation III: Foster active learning
Group problem solving and discussion
Lab exercises
Demonstrations based on class-generated data
Written and oral presentations
Projects, either group or individual

Mathematics and Statistics

What is the relationship between mathematics and statistics? Some scholars argue that mathematics is a pure theoretical science while statistics is an applied science. In their essay on the tension between mathematics and statistics, Moore and Cobb (2000: 615) present the following working hypothesis "statistics has cultural strength that might greatly assist mathematics, while mathematics has organizational strengths that provide shelter for academic statistics, shelter that may be essential for its survival."

Cobb and Moore (1997: 801) capture some of the fundamental differences between mathematics and statistics when they state that:

> Statistics is a methodological discipline. It exists not for itself but rather to offer to other fields of study a coherent set of ideas and tools for dealing with data. The need for such a discipline arises from *the omnipresence of variability*. Individuals vary. Repeated measurements of the same individual vary [...]. The focus on variability naturally gives statistics a particular content that sets it apart from mathematics. Statistics represents a different *kind* of thinking, because *data are not just numbers, they are numbers in context.* (emphasis in original)

Cobb and Moore (1997: 803) further note the distinctions between mathematicians and statisticians in the following way:

> Although mathematicians often rely on applied context both for motivation and as a source of problems for research, the ultimate focus in mathematical thinking is on abstract patterns: the context is part of the irrelevant detail that must be boiled off over the flame of abstraction in order to reveal the previously hidden crystal of pure structure. *In mathematics, context obscures structure*. Like mathematicians, data analysts also look for patterns, but ultimately, in data analysis whether the patterns have meaning, and whether they have value, depends on how the threads of those patterns interweave with the complementary threads of the story line. *In data analysis, context provides meaning.* (emphasis in original)

Moore and Cobb (2000: 623) point out that "[s]tatistics [...] values mathematical understanding as a means to an end, not as an end in itself [. . .] statistics has a subject matter of its own, quite apart from mathematics. These same statisticians argue for a cooperative synergy between the two disciplines for the following reasons:

- Despite intellectual differences, mathematics and statistics both depend on the process of working from the concrete to the abstract, and can learn from each other's successes and failures in teaching this process to undergraduates.
- Statistics can benefit from embracing more openly the importance of mathematical thinking (Moore and Cobb 2000: 625–626).

Mathematics Anxiety and Statistics Anxiety

What is mathematics anxiety and what is statistics anxiety? Phobic reactions to the study of mathematics and statistics are so common that many scholars at various universities have written about this topic and they have provided useful suggestions for overcoming this fear. McCrone (2002: 266) labels it "dyscalculia", i.e. a neurological deficit. Rossman (2006) notes that this type of anxiety occurs in elementary education. In terms of math anxiety, Iossi (2013) cites Richardson and Suinn (1972: 551, Richardson and Woolfolk 1980), who described the phenomenon of math phobia in the following way nearly half a century ago "[m]athematics anxiety involves feelings of tension and anxiety that interfere with the manipulation of numbers and the solving of mathematical problems in a wide variety of ordinary life and academic situations." That now classic article also proposed a Mathematical Anxiety Rating Scale (MARS). In this same vein, Perry (2004) notes that as many as 85% of students enrolled in introductory math classes experience some math phobia. Iossi (2013: 30–31, Bradley 2010) describes some of the strategies for dealing with math phobia. First, there are curricular approaches (retesting, self-paced learning, distance education, single-sex classes, and math anxiety courses). Second, there are instructional approaches (manipulatives, technology, self-regulation techniques, and communication). Finally, there are non-instructional approaches (relaxation therapy, and psychological treatment). To be sure, awareness of math anxiety has received academic attention for at least half a century. The recognition of this phenomenon has resulted in a wide variety of techniques and strategies to address this paranoiac reaction to the study of mathematics.

Statistics also produces anxiety in students, and there is a significant number of studies about this type of angst among students. Nearly thirty years ago, Zeidner (1991: 319) offered the following definition of statistics anxiety:

> Statistics anxiety may be construed as a particular form of performance anxiety characterized by extensive worry, intrusive thoughts, mental disorganization, tension, and physiological arousal. Statistics anxiety arises in people who when exposed to statistics content problems, instructional situations, or evaluative contexts, and is commonly claimed to debilitate performance in a wide variety of academic situations by interfering with the manipulation of statistics data and solution of statistics problems.

Zeidner (1991: 321) developed a Statistics Anxiety Inventory (SAI) to determine issues in four areas: (1) statistical procedures or activities, (2) solving of quantitative problems, (3) situations related to the study of statistics (enrolling in a statistics class, picking up a statistics textbook), and (4) evaluation of performance in statistics (exams, quizzes, and studying for a test in statistics. The content of the SAI sought to elicit information about content and performance in statistics. It was patterned after the MARS instrument (Richardson and Suinn 1972; Richardson and Woolfolk 1980). In his concluding remarks, Zeidner (1991: 327) points out that prior negative experiences with math and a low sense of math-efficacy are antecedents to statistics anxiety. Onwuegbuzie et al. (1997) also provide a detailed examination of statistics anxiety, which offers instructors useful guidelines for dealing with this phobia.

Pan and Tang's (2005: 212–214) list of references and Chew and Dillon's (2014: 205–208) bibliography attest to an anxiety-ridden response to this subject by students enrolled in statistics classes. Pan and Tang (2005: 205) employ Onwuegbuzie, DaRos, and Ryan's (1997) definition of statistics anxiety to describe this experience, namely, "anxiety that occurs as a result of encountering statistics in any form at any level." Pan and Tang (2005: 209) note that there are several factors that lead to statistics anxiety (math phobia, lack of connection to daily life, pace of instruction, instructor's attitude). However, several instructional strategies offer practical ways to address statistics phobia (practical application, real-world example carried through, orientation prior to class, multiple evaluation criteria, flexible availability of assistance). Chew and Dillon (2014: 196) highlight the significance of statistics in daily life by citing Wallman's (1993: 1) call for statistical literacy which she defines as "the ability to understand and critically evaluate statistical results that permeate our daily lives–coupled with the ability to appreciate the contributions that statistical thinking can make in public and private, professional and personal decisions".

Chew and Dillon (2014: 197) hasten to point out that statistics anxiety is not the same as mathematical anxiety. It was widely assumed that they are similar because they are related areas of study. This commonly held belief changed when Cruise, Cash and Bolton (1985) developed their Statistical Anxiety Rating Scale (STARS) to account for the differences between mathematics and statistics and their distinct types of anxiety. Subsequently, Chew and Dillon (2014: 229) provided the following definition of statistics anxiety as:

> a negative state of emotional arousal experienced by individuals as a result of encountering statistics in any form and at any level; this emotional state is preceding by negative attitudes towards statistics and is related to but distinct from mathematics anxiety.

Ultimately, there is a distinction to be made between mathematics anxiety and statistics anxiety. Nevertheless, both academic domains can create discomfort and distress in a significant portion of students who take course work in both subjects. For this reason, it is important to find ways to reduce, or, ideally, eliminate this negative reaction to both areas of inquiry.

One source of math anxiety is its use of language. Freeman and her associates Freeman et al. 2016: 283–284) make the following observations about the differences between formal writing in mathematics and ordinary language:

> Formal writing in mathematics is a precise language that requires accuracy in its expression, especially at higher levels of mathematics study [...], though it also constitutes a large part of K-12 education: in the classroom, in textbooks, and on assessments. The language of mathematics contains mathematical statements (hypotheses, conjectures, axioms, and theorems), linguistic forms and properties, grammar (connectors, combinators), and symbols. This language is often information-dense and abstract [...]. It is also vastly different than language used in social conversation [...], as is the vocabulary of mathematics with mathematical meanings being much more exact and nuanced than their ordinary definitions.

The above statement is equally applicable to the distinct nature of statistical writing and ordinary language, which may lead to statistics anxiety.

One way to accomplish this goal is to make use of what Danesi (2016: 82) calls the "Wall-less" Classroom, i.e.:

> The classroom today is becoming more and more one without walls...It is both individualist (book-based) and communal (social media-based). It thus amplifies learning and retrieves previous modes of pedagogy. This has concrete pedagogical implications, the most important one being that social media are the means through which the walls are being taken down.

Danesi (2016: 83) illustrates how social media are breaking down the classroom walls, namely, by.

1. setting homework assignments or clarifying them outside the classroom
2. exchanging ideas and solutions to classroom problems and tests
3. informing the classroom community of relevant events, such as math competitions
4. writing actual lessons for specific topics that can be shared broadly and modified according to responses–constituting an extended PolyMath Project (the project originated in a blog by Timothy Bowers in 2009 under the pseudonym D. H. J. Polymath to address unsolved mathematical problems through massive collaboration on the Internet; Nielson 2011) applied to math education.

Information

What is information? The *Merriam-Webster Dictionary* (2020) defines it as "knowledge obtained from investigation, study, or instruction." That same dictionary defines it as "a quantitative measure of information."

Gibson (2013: 349) provides a superb overview of the notion of information when she states that:

> *Information* is a concept with ancient roots that translates across multiple fields of inquiry. Use of a general model of information allows scholars to share ideas and describe information phenomena across the spectrum of academic disciplines. Information has often been defined in relation to distinct but related concepts: data, facts, knowledge, and intelligence. Information is organized data presented context, a coherent collection of messages or cues structured in a way that has meaning or use for human beings. Data may be described as a set of discrete, objective facts about events, that become information when assigned meaning or value. Facts involve information that is true, that actually exists, or can be verified according to an established standard of evaluation. Knowledge can be seen as information in context, together with an understanding of how to use that information; it is a mix of information, experience, and values that provides a framework for assessing and incorporating new information. Knowledge can be either explicit (a person is able to make this information available for introspection) or tacit (the person is not able to make the information available for introspection). Intelligence refers to the quality of the information (e.g., information concerning crucial facts, military intelligence, a secret) or the capacity of a sentient being to combine data, facts, information, and knowledge with insight and acuity. Information may therefore be defined as facts and data organized to describe a particular situation or problem and information is what people share with each other when they communicate. (emphasis in original)

In his discussion of "the information society," Danesi (2013: 354, emphasis in original) points out that this term "is used in cyberculture and media studies to refer to the economic system based primarily on the retrieval, processing, and management of information, in opposition to an economic system based on the production of the production of material goods. The latter is known as an *industrial society*." Danesi (2016: 103) further notes that the concept that today's world is the first information society is false, i.e., as he points out, every age is an information society. The dissemination of information in the past has assumed different formats and modes of distribution, e.g., oral communication in non-literate societies, hand written texts, and multiple copies of document with the invention of the printing press. Today, however, information may be stored electronically, thereby, allowing individuals access to and manipulation of vast stores of information through the use of computer programs designed to make those data available to scientists and others to meaning to these vast stores of knowledge. The electronic storage of information or data now facilitates research and teaching by making available enormous amounts of mathematical and statistical facts that can be searched with various computer programs to ascertain trends or glean meaningful information that enhances our knowledge of the world. The ease of access to these data facilitates the ever-expanding wall-less mathematical and statistical classroom.

Pop Culture and Its Relationship to Mathematics and Statistics

What is pop culture? Danesi (2019c: 4) makes the following observations about pop culture:

> There is little doubt that pop culture trends, like commodities, have fleeting value. But it is also true that pop culture constitutes an open social forum in which creativity can be expressed and displayed by virtually anyone. It is empowering, allowing common people to laugh at themselves, to gain recreation through music, dance, stories, and other forms of expression. Before the advent of pop culture as a mass form of entertainment, people sought recreational outlets through carnivals and various other public spectacles, which have typically existed alongside religious feasts since at least the medieval period. Pop culture is also a source of recreation that appeals to our fun-loving side. It is thus a modern-day descendant of carnivals. Admittedly, as most pop culture critics have suggested, most pop culture is a commodity culture. It takes place in a marketplace that is at once economic and artistic, thus appearing in short-lived and era-specific forms.

Danesi (2019c) provides specific examples of popular culture that include print culture (books, newspapers, magazines, comic books), radio culture (radio broadcasting, talk shows, Internet radio), music culture (pop music, rock and roll, hip-hop, independent music), cinema and video culture (motion pictures, video games, HBO®, Netflix®, Hulu®), television culture (sitcoms, reality TV, web TV), advertising culture (ad campaigns, placement, advertising art), pop language culture (slang, spelling style, emoji), and on line pop culture (YouTube®, Facebook®, Twitter®, memes [Danesi 2019b]).

It should also be noted that Danesi has written extensively about puzzles and problem-solving, which are forms of recreational mathematics. In one of his earliest books on this topic (Danesi 2002: ix), he talked about the "puzzle instinct," which he describes as a specific trait of *homo sapiens*—a unique propensity to solve puzzles, brain-teasers, and enigmas for the sheer delight in finding a solution. Danesi

(2019a: 115) goes on to state that "[p]erhaps in no other area of human intelligence have puzzles played as large a significant role than in mathematics." He notes that math puzzles can be traced back to the *Ahmes Papyrus*, which contains eighty-four difficult mathematical puzzles, and is one of the earliest sources of ancient mathematics. This type of "recreational mathematics" saw its popularity grow with the publication of Claude-Gaspar Bachet de Mézirac's extensive collection of math puzzles entitled *Problèmes plaisans et délectables qui se font pare les nombres* in 1612 (Bachet de Mézirac 1984). These early examples of mathematics designed to engage and amuse the public may be seen as early efforts to engage the public in the wonders of the science of math.

To this list, I would add the mathematics and statistics inherent in competitive sports. According to Sports in the United States (2020), the three most popular sports in the U.S. in terms of revenue production are: (1) American football (National Football League), (2) basketball (National Basketball Association), and (3) baseball (Major League Baseball). Other popular sports in the US are: Ice hockey, soccer, tennis, golf, wrestling, auto racing, arena football, lacrosse, box lacrosse, and volleyball. The top three professional sports (American football, basketball, major league ball) permeate every aspect of pop culture. We may view these games in person at the various stadia across North America. We listen to them on radio or on the Internet. We purchase sports apparel with the name and number of our favorite player from our favorite team. We also purchase other sports memorabilia (cards, equipment with team logos). We purchase books about our favorite sports heroes and teams and their records since their beginnings. Because of the popularity of sports in North American society by young and old alike, mathematical and statistical activities, exercises, and problems appeal to students enrolled in these classes. In fact, many sports have video games associated with them so that the fan can engage in virtual realistic competition similar to that of the professional athletes in many sports.

Games

What are the essential elements of a game? Palmer and Rodgers (1983: 3) note that games have the following components:

1. Games are *competitive*, i.e., a person competes against another individual, time, personal performance, or a goal.
2. Games are *rule-governed*, i.e., principles determine the acceptable or unacceptable actions or behavior.
3. Games are *goal-defined*, i.e., these activities have their recognized and agreed upon objectives.
4. Games have *closure*. i.e., the participants know when the activity is completed according to pre-determined criteria.

5. Games are *engaging*, i.e., these pastimes are fun and the interactants derive amusement and stimulation from engaging in them. (emphasis in original)

Sports

What is a sport? *The Cambridge English Dictionary* (2020) defines sport as "a game, competition, or similar activity, done for enjoyment or as a job, that takes physical effort and skill and is played or by following particular rules." A sport includes a wide range of games including, but not limited to, baseball, basketball, American football, ice hockey, soccer, and many similar competitive pastimes. To be sure, all of the sports just mentioned follow the principles enumerated in Palmer and Rodgers (1983: 3), i.e., they are competitive, rule-governed, goal-defined, have closure, and they are engaging. These are the attributes that appeal to fans of all sports. Sports enthusiasts love their particular teams and they engage in arguments about which team is the best. Some sports, however, allow ties, which clearly creates a problem for those who wants a definitive winner and loser. Game rules nevertheless, have established ways to address the issue of ties through total wins and losses with another team that seeks to participate in post-season playoffs in an attempt to win a championship.

Sports Wagering

What is sports wagering? According to Wikipedia (Sports Betting 2019), wagering on sports is "the activity of predicting sports results and placing a wager on the outcome." Gambling on the outcome of sports games has a long tradition. In the U.S., it may be legal, i.e., bets are placed with duly licensed companies. At this writing, nineteen states (Arkansas, Colorado, Delaware, Illinois, Indiana, Iowa, Mississippi, Montana, Nevada, New Hampshire, New Jersey, New Mexico, New York, North Carolina, Oregon, Pennsylvania, Rhode Island, Tennessee, and West Virginia) allow legal betting. Sports wagering is also legal in the District of Columbia. As a result of the incipient legalization of sports betting in the U.S., an entire industry has arisen, e.g., FanDuel®, DraftDay®, and PlayOn®.

The four major U.S. sports leagues (American football, baseball, basketball, and ice hockey) had held a position against sports gambling. However, the National Basketball Association (NBA) and Major League Ball (MLB) have advocated for a change in their previous stance against this practice. The National Hockey League (NHL) has not taken a position. At this juncture, only the National Football League (NFL) continues its oppositions to sports betting.

Sports scandals have occurred in various sports over the years, in part, because gambling was illegal. As a result, there have been various scandals caused by criminals who sought to influence the outcome of a game. One infamous case of such a crime occurred in Major League Baseball a little over a century ago. It involved the World Series of 1919 (Chicago White Sox versus the Cincinnati Reds). Because it damaged the reputation of what some have called America's sport, it came to be known as the "Black Sox Scandal" in which the White Sox deliberately lost 5 games in a nine games series. This scandal became a famous 1988 movie *Eight Men Out* based on a book by Eliot Asinof's *Eight Men Out: The Black Sox Scandal and the 1919 World Series* (1963) Another film, *The Field of Dreams* (1989) based on the novel *Shoeless Joe* by W. P. Kinsella (1982), also deals with this scandal.

Most Popular Sports in North America

What are the most popular sports in North America? According to Sports in the United States (2020), the three most popular sports in the U.S. in terms of revenue production are: (1) American football (National Football League), (2) basketball (National Basketball Association), and (3) baseball (Major League Baseball). Other popular sports in the US are: Ice hockey, soccer, tennis, golf, wrestling, auto racing, arena football, lacrosse, box lacrosse, and volleyball. The top three sports are also the most lucrative in terms of revenue through a variety of venues (ticket sales, products with specific logos, television and radio broadcast rights, and so forth).

American Football

American football has its origins in rugby football. Walter Camp (1859–1925) is widely regarded as the father of American football. Its early period, when it was developing rules and procedures, was 1869–1875. The intercollegiate period occurred from 1876 to 1893. The creation of a rules committee and athletic conferences (groups of colleges that participate regularly in intercollegiate games) dates from 1892 to 1934. The modernization of the game took place from 1933 to 1969. Modern intercollegiate football started in 1970.

Professional American football began in 1892 and various cities developed their own teams. The National Football League began in 1920 with organized and regular schedules. From 1933 to 1969, there was a period of stability and slow but steady growth. The development of a rival and competitive league, the American Football League in 1959. Ultimately, there was a merger of the two leagues in 1970, and the two leagues (NFL, AFL) compete annually in the Super Bowl to determine the winner for a given season.

Basketball

The game of basketball was invented in Springfield, Massachusetts by the Canadian James Naismith (1861–1939). He authored the first rule book for basketball on December 21, 1891. The first known intercollegiate basketball game took place on February 9, 1895 between Hamline University and Minnesota AandM. Naismith subsequently created the University of Kansas basketball team, and became its head coach (1898–1907).

The Basketball Association of American became the National Basketball Association was founded on June 6, 1946 in New York. In 1949, it merged with the National Basketball League. Subsequently, the American Basketball Association was founded in 1967. By 1976, it merged with the NBA.

Baseball

The origins of baseball derive from various games played in Europe. Immigrants brought early versions of the game to the US. In 1871, the National Association of Professional Base Ball Players was founded. In 1876, the National League was founded. In 1901, the American League came into being. Shortly thereafter, a World Series between the two leagues started in 1903. By 1905, the series became an annual event. It is known as the national sport of the US.

American Football, Basketball, and Major League Ball: Their Use as a Pedagogical Resource for Mathematical and Statistical Problems

In part II of this paper, the three most popular professional sports in North America (American football, basketball, and major league baseball) will be used to illustrate selected exemplary mathematical and statistical problems in each sport to demonstrate how these pop cultural sports manifestations may serve to provide a useful and engaging resource for teaching mathematics and statistics. All three sports (American football, basketball, and major league ball) maintain precise information about every game and every player in sports. All of these data are easily available on the Internet, so access is at one's finger tips. The examples provided in the following three sections involve the arithmetical functions of addition, multiplication, and division. To be sure, these are simple operations. Nevertheless, more complex mathematical procedures are required for certain types of information gleaned from the three sports discussed in this paper.

American Football

The famed Chicago Bears Super Bowl XX team is considered one of the best National Football squads of all time. They dominated in every area. That team produced five NFL Hall of Fame players (Richard Dent, Dan Hampton, Walter Payton, Mike Singletary, and head coach Mike Ditka). The Super Bowl itself was played in New Orleans, Louisiana at the Louisiana Superdome on January 26, 1986. Their opponent was the New England Patriots. The final score was 46–10. The time of possession (TOP) of the football for Chicago was 39.15 min while New England had the ball for 20.45 min (Super Bowl XX, 2020). This statistic is significant because it means that the team with the greatest time of possession has the best opportunity to score points and achieve more points than the opposing team.

A regulation football game lasts 1 h (60 min). One minute has 60 s. The simple formula for determining the total number of seconds in a regulation football game is:

$$60 \min \times 60 \, s = 3600 \, s$$

In Super Bowl XX, the Bears had possession of the football for 39.15 min. To determine the total number of seconds, it is necessary to multiply 39 × 60 equals 2340 s and add 15 s. The total time of possession was 2355. Based on that equation, the New England Patriots had possession for 20.45 min. The same arithmetic operation involves multiplying 20 min × 60, which equals 1200 s. Then the additional 45 s must be added for the total (= 1245 s). The final step to determine total percentage of time of possession by the Bears is to divide 3600 by 2340, which equals 65.41666%. The New England Patriots had possession for 1245 s. By using the same arithmetical operations, the Patriots had possession of the ball 34.58333% of the time. This is a substantial percentage, and it explains the lopsided final score (46–10). Three arithmetical calculations are necessary to determine the per cent time of possession of the football by the Bears in Super Bowl XX: addition, multiplication, and division.

Basketball

Michael Jordan, known by his initials "MJ" and his nickname "Air Jordan", played for the Chicago Bulls from 1984 to 1993 and 1995–1998. After a three-year retirement, he played two more years for the Washington Wizards (2001–2003). He played in a total of 1072 games over 15 seasons. During his career, he scored 32,292 points (Jordan 2020). Many people consider him to be the greatest basketball player of all time. This assertion, of course, can lead to disputes among those who believe that another player is the best. These disagreements can be resolved through mathematics and statistics. In terms of scoring, Michael Jordan's average was 30.123134 per game. The next best in that category was Wilt Chamberlain, whose nickname was "Wilt the Stilt" because of his height (7 feet, 1 inch). Jordan was a scoring guard and a small forward, while Chamberlain was a center. He played 16 seasons for several different teams including his final five seasons for the Los Angeles Lakers. He participated in 1045 games during his career with a point total of 31,419. His scoring average was 30.066028 rounded out to the next highest number, i.e., 31.1 (Chamberlain 2020).

To determine the scoring average of a basketball player, the following formula must be used:

$$SA(\text{Scoring Average}) = PTS(\text{Total Career Points}) \div G(\text{Total Games Played})$$

Arriving at this particular statistic requires two arithmetical operations: addition and division. First, it is necessary to add up all of the games played by each player. Then, that number must be divided by the total number of points scored by each player. Because the scores of each player are rounded off to the next highest number, it would appear that each player had the same number of points (30.1) over their respective careers. However, if the two players are compared numerically through arithmetic, it is clear that Michael Jordan had a slightly better scoring percentage based on the formula use to determine this type of statistical information.

The use of scoring average, however, is just one way to compare two players. Thus, it is possible to compare two players based on very specific aspects of the game, e.g., field goals per game (FG), field goal attempts (FGA), field goal percentage (FG%), three point goals per game (3P), three point field goal attempts per game (3P%), two point field goals per game (2P), two point attempts per game (2PA), two point percentage (2P%), effective field goal percentage (eFG%), free throws per game (FT), free throw attempts per game (FTA), free throw percentage (FT%), offensive rebounds per game (ORB), defensive rebounds per game (DRB), total rebounds per game (TRB), assists per game (AST), steals per game (STL), blocks per game (BLK), turnovers per game (TOV), personal fouls per game (PF), and points per game (PTS). Thus, a comparison of two players can result in multiple statistics. Each of these statistics requires a formula to determine with precision a given player's actual performance based on at least two dozen parameters. Each one of these provides subtle insights into an individual's strengths and weakness. As a result, students of mathematics and statistics can fine tune their arithmetic skills as well as make arguments for and against the quality of a specific player.

Major League Baseball

Baseball aficionados love to assess their favorite players by comparing their numbers or stats. This type of information includes batting average, homeruns scored, runs batted in, runs scored, stolen bases. While it is quite easy to memorize these bits of information. A determination of the mathematical and statistical processes employed to arrive at these data requires some knowledge of the procedures required to arrive at these facts and figures (Martin and Guengerich 2004; Ross 2007).

Frank Thomas (1968-), whose nickname was "The Big Hurt", played for the Chicago White Sox for most of his career (1990–2005) and for two other teams during his last three seasons (Thomas 2020). He was elected to the Baseball Hall of Fame on his initial year of eligibility in 2014. His batting average and his slugging percentage provide an excellent way to teach some basic arithmetic concepts, namely, addition and division. To determine a baseball player's batting average, the following formula is used:

$$\text{Batting Average} = \left(\text{Hits} \div \text{At Bats}\right)$$

Frank Thomas had 8199 at bats and 2468 hits. It is necessary to divide his total number of at bats by his total hits. This produces a batting average of .3010123. On the other hand, to calculate a baseball player's slugging percentage (SLG) involves the following formula: TB (Total Bases) = 1B (First Base) + 2 × 2B (Second Base) + 3 × 3B (Third Base) + 4 × HR (Home Runs). This may be formulated as follows. The World Wide Web now has a Slugging Percentage Calculator (2020), but its use would not allow a student to engage in the necessary mathematical calculations to internalize the procedures to carry out the SLG.

$$\text{SLG} = \frac{\left(1\text{B}\right)+\left(2\times 2\text{B}\right)+\left(3\times 3\text{B}\right)+\left(4\times \text{HR}\right)}{\text{AB}}$$

The translation of this formula is Total Number of Bases = 1B (the number of singles) + 2 × 2B (the number of doubles) + 3 × 3B (the number of triples) + 4 × HR (the number of home runs). Thus, the slugging percentage (SLG) is as follows.

$$\text{SLG} = \left(\text{TB} \div \text{AB}\right).$$

In the case of Frank Thomas, he had 2468 total bases during his career. Likewise, he had 1440 singles, 495 doubles, 12 triples, and 521 home runs. It is necessary to multiply the singles by 1 (= 1440) + two times the number of doubles (= 990) + three times the number of triples (= 36) + four times the number of home runs (= 2084). This totals 4550. Next, this sum must be divided by the total number of at bats (= 8199). The slugging percentage (SLG) is .5549457. The SLG measures the quality of a batter's hits, while the batting average measures the number of times on base. Frank Thomas's SLG of .554 is outstanding.

Frank Thomas's batting average and slugging percentage represent only a small part of any baseball player's total skill. Other factors include runs batted in (RBI), stolen bases (SB), caught stealing (CS), bases on bats/walks (BB), strikeouts (SO), times hit by a pitch (HPB), sacrifice hits/bunts (SH), sacrifice flies (SF), and intentional bases on balls (IBB). All of these game dynamics play a part in the assessment of the total value of an individual baseball player. The arithmetical computations needed to ascertain Frank Thomas's batting average requires a knowledge of addition and division. Likewise, his slugging percentage requires a knowledge of addition, multiplication, and division. Danesi (2008: 40–77) offers a very useful semiotic perspective to teach these basic arithmetical operations.

Furthermore, the example of selected numbers related to Frank Thomas's batting performance demonstrates how baseball managers make decisions based on statistical information. In the cases of hits and at bats, batting average is less important than slugging percentage because the latter clearly measures the quality of the hits versus the number of on base hits made. It must be remembered that a winning team in baseball must have more home runs than the opponent. This fact accounts for the significance of certain statistics in this sport.

Concluding Remarks

This chapter has addressed the significance of a popular cultural phenomenon (professional sports) and how they can be used to engage students in learning about mathematics and statistics by means of bringing mathematics and statistics into the world and bringing the world into the classroom. Danesi (2016: 137) points out that the "wall-less classroom [...] can now be defined not as a replacement of, but as an extension of, the traditional classroom–that is why the critical components of the latter are still in the picture (so to speak). The main feature of education is still the teacher-student relationship."

The first part provided definitions and discussions of key concepts (mathematics, mathematics pedagogy, statistics, statistics pedagogy, mathematics and statistics, mathematics anxiety and statistics anxiety, information, pop culture and its relationship to mathematics and statistics, games, sports, sports wagering, and sports in North America). The second part provided selected examples of the use of mathematics and statistics to in the three most popular sports in North America (American football, basketball, major league baseball). Particular examples from these sports demonstrated how the use of these popular cultural manifestations can teach students how to apply basic mathematical and statistical notions from real world data located on the Internet. The latter is a cornucopia of information, i.e., big data, which, in sports provides reliable data gathered over a player's career.

References

Asinof, E. (1963). *Eight men out: The black sox scandal and the 1919 world series*. New York: Holt, Rinehart and Winston.

Bachet de Mézirac, C. J. (1984 [1612]). *Problèmes plaisans et délectables qui se font par les nombres*. Lyons: Gauthier-Villars.

Berggren, J. L., Gray, J. J., Knorr, W. L. R., Fraser, C. G., Folkerts, M. (2020). Mathematics. *Encyclopedia Britannica*. Retrieved from https://www.britannica.com/science/mathematics.

Bradley, L. M. (2010). Working with adults with math phobia. Seminar paper. University of Wisconsin, Platteville. Retrieved from https://minds.wisconsin.edu/bitstream/han-dle/1793/46835/BradleyLouise.pdf?sequence=4andisAllowed=y.

Cambridge English Dictionary (2020). Sport. Retrieved from https://dictionary.cambridge.org/us/dictionary/english/sport.

Chamberlain, W. (2020). Retrieved from https://www.basketball-reference.com/players/c/chambwi01.html.

Chew, P. K. H. and Dillon, D. B. (2014). Statistics anxiety update: refining the construct and rec-ommendation for a new research agenda. *Perspectives on Psychological Science* 9: 196-208.

Cobb, G. (1992). Teaching statistics. In: L. A. Steen (ed.), *Heeding the call for change: Suggestions for curricular action*, 3-43. Washington, DC: Mathematical Association of America.

Cobb, G. W. and Moore, D. S. (1997). Mathematics, statistics, and teaching. *The American Mathematical Monthly*, 104(9), 801-823.

Cruise, R. J., Cash, R. W., and Bolton, D. L. (1985). Development and validation of the instrument to measure statistical anxiety. *American Statistical Association 1985 proceedings on the sec-tion on statistical education*, 92-97. Washington, DC: American Statistical Association.

Danesi, M. (2002). *The puzzle instinct: The meaning of puzzles in human life.* Bloomington: Indiana University Press.

Danesi, M. (2008). *Problem-solving in mathematics. A semiotic perspective for educators and teachers.* New York: Lang Publishing.

Danesi, M. (2013). Information society. In: M. Danesi (ed.). *Encyclopedia of media and commu-nication*, 354-356. Toronto: University of Toronto Press.

Danesi, M. (2016). *Learning and teaching mathematics in the global village: Math education in the digital era.* Basel: Springer Nature.

Danesi, M. (2019a). *An anthropology of puzzles. The role of puzzles in the origins and evolution of mind and culture.* London: Bloomsbury Academic.

Danesi, M. (2019b). Memes and the future of popular culture. *Popular Culture*, 1(1), 1-81.

Danesi, M. (2019c). *Pop culture: Introductory perspectives.* 4th ed. Lanham: Rowman and Littlefield.

Devlin, K. (2000). *The math gene: How mathematical thinking evolved and why numbers are like gossip.* New York: Basic Books.

Freeman, B., Higgins, K., and Horney, M. (2016). How students communicate mathematical ideas: An examination of multimodal writing using digital technologies. *Contemporary Educational Terminology* 7: 281-313.

Gibson, T. (2013). Information. In: M. Danesi (ed.). *Encyclopedia of media and communication*, 349-354. Toronto: University of Toronto Press.

Iossi, L. (2013). Strategies for reducing math anxiety in post-secondary students. In: S. M. Nielsen, and M. S. Plakhotnik (eds.), *Proceedings of the sixth annual college Education research con-ference: Urban and international section*, 30-35. Miami: Florida International University. http://coeweb.fiu.edu/rsearch_conference/

Jordan, M. (2020). Retrieved from https://www.basketball-reference.com/players/j/jordami01.html.

Kinsella, W. P. (1982). *Shoeless Joe.* New York: Houghton Mifflin.

Lexico (2020). Statistics. Retrieved from https://www.lexico.com/en/definition/statistics.

McCrone, J. (2002). Dyscalculia. *The Lancet of Neurology* 1: 266.

McLuhan, M. (1962). *The Gutenberg galaxy: The making of typographic man.* Toronto: University of Toronto Press.

Martin, H. and Guengerich, S. (2004). *Integrating math in the real world: The math of sports.* Portland, Maine: J. Weston Walch Publisher.

Merriam-Webster Dictionary (2020). Information. Retrieved from https://www.merriam-webster.com/dictionary/information.

Moore, D. S. and Cobb, G. W. (2000). Statistics and mathematics: Tension and cooperation. *The American Mathematics Monthly* 107: 615-630.

Nielson, M. (2011). *Reinventing discovery: The new era of networked science*. Princeton: Princeton University Press.

Nuessel, F. (2012). The representation of mathematics in the media. In: M. Bockarova, M. Danesi, and R. Núñez (eds.) *Semiotic and cognitive science essays on the nature of mathematics*, 165-208. Munich: Lincom Europa.

Onwuegbuzie, A. J., DaRos, D., and Ryan, J. (1997). The components of statistics of statistics anxiety: A phenomenological study. *Focus on Learning Problems in Mathematics* 19: 11-35.

Palmer, A. and Rodgers, T. S. (1983) Games in language teaching. *Language Teaching* 16: 2-21.

Pan, W. and Tang, M. (2005). Students' perceptions on factors of statistics anxiety and instructional strategies. *Journal of Instructional Psychology* 32: 205-214.

Perry, A. B. (2004). Decreasing math anxiety in college students. *College Student Journal* 38: 321-324.

Richardson, F. C. and Suinn, R. M. (1972). The mathematics anxiety rating scale: Psychometric data. *Journal of Counseling Psychology* 19: 551-554.

Richardson, F. C. and Woolfolk, R. I. (1980). Mathematics anxiety. In: I. G. Sarason (ed.), *Test Anxiety: Theory, research, and applications*, 271-288. Hillsdale: Lawrence Earlbaum.

Rossman, S. (2006). Overcoming math anxiety. *Mathitudes*, 1: 1-4.

Ross, K. (2007). *A mathematician at the ballpark: Odds and probabilities for baseball fans*. New York: Plume.

Slugging Percentage Calculator (2020). Retrieved https://miniwebtool.com/slugging-percentage-calculator/B

Sports betting. (2019). Retrieved from https://en.wikipedia.org/wiki/Sports_betting.

Sports in the United States. Retrieved from https://en.wikipedia.org/wiki/Sports_in_the_United_States.

Super Bowl XX (2020). Retrieved from http://www.rauzulusstreet.com/football/superbowl/superbowlXX.htm.

Thomas, F. (2020). Retrieved from https://www.baseball-reference.com/players/t/thomafr04.shtml.

Wallman, K. K. (1993). Enhancing statistical literacy: enriching our society. *Journal of the American Statistical Association* 88 (421): 1-8.

Williams, T. J., Anderson, D. R., and Sweeney, D. J. (2020). Statistics. *Encyclopedia Britannica*. Retrieved from https://www.britannica.com/science/statistics

Zeidner, M. (1991). Statistics and mathematics anxiety in social science students: some interesting parallels. *British Journal of Educational Psychology* 61: 319-328.

Chapter 7
Travels with Epsilon in Sign and Space

Louis H. Kauffman

Introduction

This paper is about the relationship of diagrams with mathematics.

Mathematics is replete with diagrams of all kinds such as the classical diagrams of Euclidean Geometry and the wilder diagrams of topology. Indeed, symbolisms in mathematics such as the Leibniz notations for integration and differentiation are themselves diagrams indicating the very processes that they represent.

$$\int_{0}^{x} f(t)\,dt = F(x)$$
$$dF(x)/dx = f(x)$$

What is less obvious is how certain forms can exhibit shape that links different areas of mathematics via a common structure that lives in the diagrams.

We study the linking of mathematical fields in this paper by examining first a magical diagrammatic for vector calculus, and then showing how it works and why it works by relating that formalism to the question of coloring maps and graphs in and out of the plane. In the course of this journey we shall have a trivalent vertex that we call the *epsilon*. Different ways of viewing the way the epsilon works and behaves shed light on the structure of dot products of vectors, cross products of vectors, multiple cross products, the structure of the quaternions and edge coloring problems for graphs that are equivalent to the Four Color Theorem. Once one has

L. H. Kauffman (✉)
Department of Mathematics, Statistics and Computer Science, University of Illinois at Chicago, Chicago, IL, USA

Department of Mechanics and Mathematics, Novosibirsk State University, Novosibirsk, Russia
e-mail: kauffman@uic.edu

© Springer Nature Switzerland AG 2020
S. A. Costa et al. (eds.), *Mathematics (Education) in the Information Age*, Mathematics in Mind, https://doi.org/10.1007/978-3-030-59177-9_7

taken this journey, neither graphs and colorings nor vectors and their algebra are ever the same again. It all pivots on the epsilon.

The second journey in this paper is into diagrammatic knot theory. There we show how the diagrams of knot theory, decorated shadows of projections from three dimensions, are intimately related to non-associative algebras called quandles. The simplest quandle involves three colors and is, in its structure, very close to the coloring problems we have considered earlier in the paper. But now the associations are with topology and how the algebra helps uncover hidden topological properties.

The third journey examines the resolution of a diagrammatic singularity and finds a generalized epsilon and the Jacobi identity for Lie algebras hidden in the diagrams.

A longer tale can be told here, but we hope that this introduction to the ways of diagrams gives the reader a taste of this way to imagine the roots of mathematics.

In the first part of the paper. The author is in dialogue wth a fictional mathematician named RosePen. Professor RosePen is a figment of the author's imagination, influenced by the ideas, discoveries and inventions of Roger Penrose, John H Conway, George Spencer-Brown, Charles Sanders Peirce, Lewis Carroll and other great contributors to the diagrammatic interfaces in the making of sign and space.

Acknowledgement Kauffman's work was supported by the Laboratory of Topology and Dynamics, Novosibirsk State University (contract no. 14.Y26.31.0025 with the Ministry of Education and Science of the Russian Federation.)

A Magic Calculus of Vectors

I went to the CMF last year. That's the Convention on Mathematical Fictions. Sometimes we call it the CFM, the Convention on Fictional Mathematics. Well, call it what you will, we were still meeting in person then and sitting down to scraps of paper and scribbling funny geometries and strange equations. You remember how it was. And I met this guy RosePen and he sits me down and says. Look. You have to learn my graphical rules. They will change your life. I says - yeah, really? And he says Really! So I sat down at the bar with him in the Atlanta Ritz Carlton and he takes out a sheet of paper, a bit crumpled.

He makes the drawing you see in the figure below, and he says this is a vector.

I says, it looks like a blob with a line hanging on it. Yep! He says. That's a vector. And here is the dot product of two vectors. He draws two of his vectors and joins their arcs together like this .

$$\text{\small ⒜ ⒝} = A \cdot B$$

I says. Hmm... I guess you are going to tell me that if a blob has no hanging strings, then it is a scalar? Right! He says. How did you know? I says, look you told me that thing there is a scalar product (dot product) and so I figured you joined those arcs to get rid of them. Well. He says. You are absolutely right! Can you figure out what would be the vector cross product?

Aw, I says. Well, you have to combine two vectors to get a vector. I gather your vectors just have one arc attached to the blob. So I wager you need a trivalent node like this

$$\text{\small ⒜ ⒝} = A \times B$$

and you can run the arcs from your vector into two of the three lines on the tri-node and you will have a new blob with one arc! That's my guess for the cross product.

I couldn't help myself. I continued. I says: Look. You are gonna have to have that.

A x A is zero and that A x B is perpendicular to A and to B. And you are gonna need that A x B = − B x A. So there is a lot of work to be done here. I think we better start with A x B = − B x A. This is what you need!

$$\text{\small ⒜ ⒝ ⒜ ⒝} = -$$

That twisted thing is B x A and you really need your trivalent vertex to satisfy the same identity!!

$$\gamma = -\gamma$$

Twist two legs of that trivalent vertex of yours and thing changes sign. Now its ok because we will have A x A = − A x A and so A x A = 0. No sweat!

He looks at me with slitted eyes, a bit suspicious you know. And he says. You are exactly right. Nobody ever got this before. Are you from the CIA? Maybe I should just stop talking right here. Naw, I says. I never talk to the Cantorian Infinite Adepts. They are too theological for me.

But look, I says, your system works too well! Look at (A x B).C where I use a period for the dot product. We get a clear proof that

$$(A \times B).C = A.(B \times C)$$

by just deforming your diagrams. Ha!

$$
\begin{array}{cc}
(A \times B) \cdot C & A \cdot (B \times C) \\
\| & \| \\
A \ B \ C & \simeq A \ B \ C
\end{array}
$$

You catch on fast, he says. But now I will tell you the *secret*. We call the trivalent node our *epsilon*. And here is the *epsilon identity*.

$$\big\rangle\!\big\langle = -\,)\,(\;+\;\big\rangle\!\big\langle$$

I shall initiate you into its vectorial secrets. You mean, I says, you can derive other identities from this secret identity. He smiled. A co-conspirator, I thought. Well, I decided to play along. So I says, Ok wise-guy lets try the notorious. Vector Triple Product: (A x B) x C. What will your smart diagrams do with this stumper? Here, don't tell me. I'll do it! There it is.

$$
\begin{array}{c}
(A \times B) \times C \\
= A \ B \ C
\end{array}
$$

I look and look at this. And then I remember his epsilon identity and it re-forms in my mind, slightly deformed:

And I say, why not! You told me I could do this and I know you don't care if I deform it a little. Now I will put the blobs back on top. Aha! There it is. Yes!

$$(A \times B) \times C = -A(B \cdot C) + (A \cdot C)B$$

I put the blobs back and the epsilon identity became that familiar formula

$$(A \times B) \times C = -A(B.C) + (A.C)B$$

from our beloved vector calculus. And I says to RosePen. What the heck. How did you do that? That is a complicated geometrical formula and your diagrams make it fall out of nowhere. What is going on here. Are vectors really something other than what I thought they were? What planet are you from?

Then I decided to try something simpler. I says to RosePen what about the fact that A x B is perpendicular to A and to B? Can we see that? I know. I know. You are going to say that perpendicularity of V and W is defined by the eq. V.W = 0. Ok. Then I am supposed to prove that (A x B). A = 0. Oh wait. I don't even need the diagram. After all I just did show that A x A = 0 for any A. So (A x B). A = − (B x A). A = − B. (A x A) = 0 and we are done. Ok I am satisfied. Lets go back to your special epsilon identity. What about A x (B x C). We can do that and find that A x (B x C) = −A (B.C) + (A.C)B.

$$A \times (B \times C) = \quad = -A\,B\,C + A\,B\,C$$

So I get this beautiful difference formula

$$(A \times B) \times C - A \times (B \times C) = -A(B.C) + (A.B)C.$$

And this shows very explicitly how the vector cross product operation is not associative.

$$A \times (B \times C) - (A \times B) \times C = -A\,B\,C + A\,B\,C$$

RosePen intervened and said. Why don't you try for associativity? Can you make an associative product from these materials? I said. Wait. I remember the definition of *quaternion multiplication*. It is

$$UV = -A.B + A \times B.$$

You almost do cross product multiplication, but you add that scalar product.
    Why don't you try it? He says.
    Ok. I will says I. I will define.

$$uV = -u \cdot V + u \times V$$

And quaternions are four dimensional vectors, so if U and V are three dimensional, then a quaternion is of the form a + V where a is a scalar. So we have

$$(a + U)(b + V) = ab + aV + bU + UV = ab + aV + bU - U.V + U \times V,$$

quite a mixture of scalar and vector products. It is no wonder that after the "quaternion wars" in the nineteenth century most applied mathematicians wanted to work separately with scalars, vectors, scalar products and vector products. But the quaternions get around, and they are really fundamental for understanding three and four dimensional space. Note that, from our formula above, we have that UU = -U.U, and so if U has length 1 we have UU = -1. We have a whole sphere's worth of square roots of minus one!

Well. In this case I won't bore you with the calculation showing that quaternion multiplication is associative. You'll see that it works out. If I, J and K are three perpendicular vectors of unit length so that II = J.J = K.K = 1, so we have

$$II = JJ = KK = IJK = -1,$$

the famous formula for the quaternions discovered by Sir William Rowan Hamilton in 1843. You know what he said about it:

> ...an under-current of thought was going on in my mind, which gave at last a result, whereof it is not too much to say that I felt at once the importance. An electric circuit seemed to close; and a spark flashhed forth, the herald (as I foresaw, immediately) of many long years to come of definitely directed thought and work, by myself if spared, and at all events on the part of others, if I should even be allowed to live long enough distinctly to communicate the discovery. Nor could I resist the impulse - unphilosophical as it may have been - to cut with a knife on a stone of Brougham Bridge, as we passed it, the fundamental formula with the symbols, i, j, k; namely, ii = jj = kk = ijk = - 1 which contains the Solution of the Problem... (Altmann 1986)

And I stopped for moment and then I said. Wow! Look at this one!!

$$(A \times B) \cdot (C \times D) = A \, B \, C \, D$$

$$= -A \, B \, C \, D + A \, B \, C \, D$$

$$= -(A \cdot B)(B \cdot C) + (A \cdot C)(B \cdot D)$$

I turned to RosePen and I said. You had better explain what is going on here.

RosePen's Explanation

In order to explain this to you, RosePen said, I have to tell you about a problem that does not seem to have anything to do with three dimensional space or vectors or dot products. The problem I am concerned about is a problem of coloring the edges of a network with trivalent nodes, using three colors: red (r), blue (b) and purple (p). It is very convenient for me to think of purple as a superposition of red and blue and so I will write p = rb and make drawings like this.

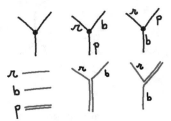

In this drawing you see that I color a red line red and a blue line blue, but I color a purple line by a combination of red and blue. The RULE for my coloring problem is that there must be three distinct colors at each node in the network. Thus at a trivalent node drawn in the plane, you will see the cyclic order of rbp or rpb, and I can make my drawings as illustrated using only red, blue and the superposition of red and blue that I call purple. Then the solution to a coloring problem looks like this.

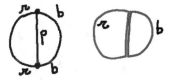

I can indicate the solution to a coloring problem by putting letters on the edges of the graph (the network), or I can color the edges. When I color them I have a collection of blue loops and red loops. The blue loops do not touch each other and the red loops do not touch each other. Red and blue loops can share segments of arc that correspond to purple edges in the network. Here is a more complex example.

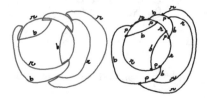

It is from this coloring problem that I conceived of the epsilon identity, for you see that the parallel and crossed arcs arise naturally when one looks at the color interactions of two nodes.

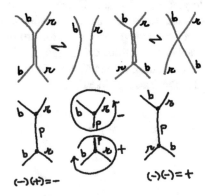

If the cyclic permuations of colors are opposite on the two nodes, then we can pull the purple superposition apart and get nearby uncrossed blue and red curves. If the cyclic permutations are the same, then we have a red arc crossing a blue arc. These are the only two structural possibilities for the color interaction of two nodes. Of course we have singled out purple for the sake of emphasis, but the same remarks would apply if the middle line were another color. So I label the case of parallel arcs with a minus sign to indicate that the two permutations are of opposite sign! And we get the epsilon identity as an expression of the coloring possiblitiies. I think of the identity in color like this:

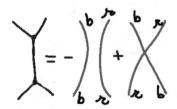

And I discovered a most remarkable fact. *If I place a square root of negative unity at each node of the network and then expand the edges by the epsilon identity I can count the number of colorings of the net.*

I says, to RosePen. If you put an i (with ii = −1) at every node, then you can just reverse the sign of the terms in the epsilon identity.

He says. Yes. That is what happens and I get a formula like this.

$$[\textstyle\bigtimes] = [)(] - [\textstyle\bigtimes]$$

$$[\textstyle\bigcirc] = 3$$

In applying this formula you erase an edge in two copies of the graph, and you replace the edge by two parallel arcs in one copy and by two crossed arcs in the other copy.

And here are two examples of color counts.

$$[\Phi] = [\text{oo}] - [\infty]$$
$$= 3^2 - 3 = 6.$$
$$[\text{o-o}] = [\text{o-o}] - [\infty]$$
$$= 3 - 3 = 0.$$

In the first case there are indeed six ways to color this graph. In the second case the graph is not colorable and the formula gives the correct answer zero. This graph is planar but it can be disconnected by removing an edge. There is a famous Theorem called the "Four Color Theorem" and it is equivalent to the statement:

Theorem *A planar trivalent graph G that cannot be disconnected by removing an edge can be colored with three colors on its edges so that every node receives three distinct colors.*

This means that the formula [G] will always be non-zero for any such graph G.

The Theorem does not have a simple proof. I am hoping that an analysis of this formula will yield a simple proof of the Theorem.

So I says. Ok. I see how you found the epsilon identity, but it is still a mystery to me what it has to do with three dimensional space. Is this some mysticism about your three colors?

RosePen replies. I had better say a few more words about coloring before we to back to vectors. Look at this diagram.

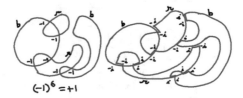

I have illustrated how each node of a coloring is assigned +i or –i according to the epsilon gives it +1 or − 1. Notice what happens when we have a red arc crossing a blue arc (or vice versa). Then one corner gets I and the the other gets –i. The product of (+i) and (−i) is (−1). So each time a red curve and a blue curve cross, my bookkeeping registers a negative one. If there is a bounce (no crossing) as shown in the figure, then we get a minus i and a plus i, so the product is one. Thus bounces contribute a + 1. Therefore if we have a coloring of a planar graph and we take the product of all my +i and −i contributions it will equal one − because curves in the plane intersect one another an even number of times (by the Jordan Curve Theorem). Here is an example for you to look at. This is why my sum [G] must always count one for each coloring of a plane trivalent graph.

Now lets look more closely at the epsilon identity. I will make definitions. I let $\varepsilon[rst]$ be a a text symbol for the epsilon node with some specific assignment of values for r,s,t from among the colors r, b and p. Then I will define

$$\varepsilon\left[rbp\right] = \varepsilon\left[bpr\right] = \varepsilon\left[prb\right] = 1$$
$$\varepsilon\left[rpb\right] = \varepsilon\left[pbr\right] = \varepsilon\left[brp\right] = -1$$

And $\varepsilon[rst] = 0$ if any two of these labels are equal to one another. These are rules we have used in coloring.

Then the epsilon identity becomes an algebraic statement about the values of the epsilon. It looks like this.

$$\sum_{t} \varepsilon_{rst}\, \varepsilon_{tuv} = -\delta_v^{\,r}\delta_u^{\,s} + \delta_u^{\,r}\delta_v^{\,s}$$

You have to stare at this formula for a while to see that it is actually very simple. The deltas are what we call Kronecker Deltas, delta[x,y] = 1 only if x = y and it is 0 otherwise. The sum on t in the formula above just amounts to taking the value different from both r and s or from t and u because our epsilon vanishes where there are equal indices and we only have three indices to work with. I will illustrate the actual cases for you below.

The upshot of this way of thinking of the epsilon identity in terms of indices is that we can interface it with vectors. What is a vector? I told you earlier to think of a vector as a blob with an arc hanging down, but the usual way to think of a vector is as a triple of numbers such as a = (a1, a2, a3). You can think of the line for the blob as a place to write the index so that for example:

$$a_3 = a_{\big)3}$$

Then the dot product follows once we use the rule that *you must sum over all the possible index values for an arc that has no free ends.*

$$a \cdot b = a_1 b_1 + a_2 b_2 + a_3 b_3$$

I hope you now see how our arc-connection diagram corresponds to the dot product. Just so, our diagram for the cross product is actually a definition for the cross product. I will calculate one component for you below and you will see that it is working!

$$a \times b = a \overset{b}{\underset{?}{\bigvee}} \qquad \overset{+}{\underset{3}{\bigvee}} = +1, \; \overset{}{\underset{3}{\bigvee}} = -1$$

$$(a \times b)_3 = a \underset{3}{\bigvee} b = a \underset{3}{\bigvee} b + a \underset{3}{\bigvee} b$$

$$(a \times b)_3 = a_1 b_2 - a_2 b_1$$

In fact, he says, you see that the epsilon gives us the determinant just like this.

$$\overset{a \quad b \quad c}{\underset{i \quad j \quad k}{\bigvee}} = \sum_{ijk} \varepsilon_{ijk} a_i b_j c_k = \begin{vmatrix} a_1 & a_2 & a_3 \\ b_1 & b_2 & b_3 \\ c_1 & c_2 & c_3 \end{vmatrix}$$

So we have that DET(a,b,c) = a.(b x c).

$$\overset{a \quad b \quad c}{\bigvee} = \overset{a \quad b \quad c}{\smile} = a \cdot (b \times c)$$

And there is a well known formula for the vector cross product that would formally put the perpendicular unit direction vectors I, J and K in the first row.

$$\begin{vmatrix} I & J & K \\ b_1 & b_2 & b_3 \\ c_1 & c_2 & c_3 \end{vmatrix} = b \times c$$

Well, I thought about that, and I worked out the other two components of the vector cross product and it was all logically clear. So we really had proved all those identities and more by just drawing topological diagrams and using the diagrammatic epsilon identity. But it still seemed to me as mysterious as ever. Why should something like this work? I had never thought of vectors as topological before. Before this conversation with RosePen, I always thought of vectors as rigid arrows that could make angles with each other and that they were the underpinning of a corresponding geometry of lines and planes and sharp directions. So I asked him more questions.

I said. Well Professor RosePen, I still do not quite understand what is going on here.

Do you mean to tell me that properties of vectors are behind the questions about coloring graphs? Or do you mean that the properties of graph colorings are the subtle structure of vectors?

He smiled and said "Yes."

An Intermediate Epilogue

I had to go. And I am still puzzling about this connection to this day. Just yesterday I ran across a paper by Kauffman (Kauffman 1990) entitled "Map Coloring and the Vector Cross Product". I could almost imagine what that might be about. I read it and continued to think about this colorful and disturbing way to look at vectors and vector calculus.

I have to tell you about this. Kauffman reformulated the coloring problem *entirely* in terms of the vector cross product! He turned it into some arcane property of perpendicularities. And I still don't understand anything! You'll see. Lets go back to I and J and K, ok? And we are looking at the cross product algebra so that I x I = J x J = K x K = 0, but I x J = K and J x I = − K and all that. It is just a way to talk about epsilon by now. But this is a weird algebra. It is not associative.

$$\left(I \times J \right) \times J = K \times J = -I,$$

$$I \times \left(J \times J \right) = I \times 0 = 0.$$

So Kauffman poses this problem. Suppose you take a product of some variables, any number of variables, like X, Y, Z, W and you associate it in two ways and write the equation stating that the result of the multiplication is the same for both associations. For example, you could write

$$\left(X \times Y \right) \times \left(Z \times W \right) = X \times \left(Y \times \left(Z \times W \right) \right).$$

Kauffman then asks you to solve this equation, using only I or J or K for the values. You get to use a value for more than one variable, but ony get to use I or J or K.

Neither side of your solution can be zero. You have to produce two equal non-zero products. Can you solve it for this example? Well in the example an answer is X = I, Y = J, Z = J, W = K. Try it! Kauffman claims to be able to solve all such equations in any number of variables.

It seems to be a tricky problem about combinations of perpendicularities. But that isn't how Kauffman solves these problems. He uses the graphical calculus. Then we have:

$$(X \times Y) \times (Z \times W) = X \ Y \ Z \ W$$

$$X \times (Y \times (Z \times W)) = X \ Y \ Z \ W$$

He puts them together in one graph by taking the mirror image of one expression and tying it to the other.

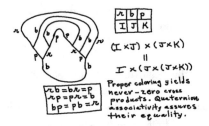

$$(I \times J) \times (J \times K)$$
$$\shortparallel$$
$$I \times (J \times (J \times K))$$

Proper coloring yields never-zero cross products. Quaternion associativity assures their equality.

Then you color the graph using r,b,p and read out a solution to the vector cross product equation by taking I for r, J for b and K for p. I kid you not. Since we have chosen a proper coloring of the graph of the two tied trees, all the partial products in the vector cross products will be non-zero. But this means that we can view these products as *quaternion products* (since in the quaternions the non-zero products of I and J and K are the same as the vector cross products). Thus the two associated products have to be equal *because the quaternions are associative*, and we are done! You can check that indeed

$$(I \times J) \times (J \times K) = I \times (J \times (J \times K)).$$

It turns out that the full coloring problem for arbitrary planar trivalent graphs is implied by the coloring of tied trees. This makes the Four Color Theorem (Appel and Haken 1977; Apprl et al. 1977; Heawood 1890) equivalent to this property of solutios to equations involving associated vector cross products. At this stage in mathematics we do not fully understand why maps can be colored (although there is a complex proof) and we do not fully understand the relationships among graphs, vector cross products and the quaternions. There is much to learn in this domain. Perhaps it will all become clear one day and we shall understand the whole story. For now, it is a fascinating ground for research. The relationship of particular mathematics with the geometry and topology of diagrams will become ever more important to the unity of mathematics and for the gesture that it makes to the unity of the world.

Knots

Diagrams are often tied to specific contexts and so the best way to indicate the wider generality behind the examples of mathematical connection that we have drawn in this paper is to give another example of the phenomenon. In this case, I want to show, how by following the diagrams one can see a deep connection between knot theory and the mathematics certain algebras. I shall be a brief as possible, and start with the knot theory. In knot theory we use diagrams like this.

The diagram represents a curve in three dimensional space that goes under and over itself in the weaving pattern of the drawing. The diagram uses the well-known drawing convention that a broken line is a projection from space such that the unbroken arc that crosses the broken part is higher than the "broken arc" that proceeds underneath. We can represent topological movenments of knots (called istopies) by changes in the diagrams. For example, view the diagram below.

It should be clear to you that the complicated curve on the left can be undone and transformed to the unknotted loop on the right. In fact there is a system of moves on the diagrams that can accomplish this aim. The basic moves are shown below. These are called the *Reidemeister Moves* after Kurt Reidemeister, who wrote the first book on the theory of knots.

Here are two examples of unknotting and unlinking using the moves.

In the first case, we use the II move and then the I move to unknot this single curve. In the second case we use a III move to simplify the rings, and then three II moves to undo them completely. The second example is interesting because it actually needs the III move to be undone.

Now I will show you a way to related algebra to these diagrams. We will have a way to "multiply" elements a and b, denoted ab. And we shall label arcs in the knot and link diagrams by these elements. When an arc a under-crosses another arc b, then the exiting arc will be labeled by the product ab as shown below.

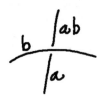

See the diagrams below.

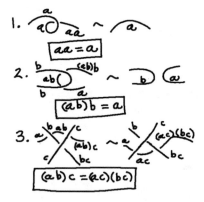

We want the labeling to respect the Reidemeister moves and this leads to algebra rules:

1. aa = a.
2. (ab)b = a
3. (ab)c = (ac)(bc).

An algebra that satisfies these rules is called a *quandle*. Here is a very simple example of a quandle. We shall have three algebra elements a,b and c. And we shall have the rules aa = a, bb = b, cc = c and ab = ba = c, ac = ca = b, bc = cb = a. In other words, any single element combines with itself to produce itself. And two distinct element combine to produce the third element. Indeed this algebra is similar to our colring rules for r,b and p but there the colors combine with themselves by different rules.

Note that (ab)c = cc = c while (ac)(bc) = ba = c. So we have (ab)c = (ac)(bc) as desired for the third Reidemeister move. You can check the other cases easily. For example, (aa)b = ab = c and (ab)(ab) = cc = c. We will use this three color algebra {a,b,c} to color knots and links! Here is a coloring of the trefoil knot.

The trefoil is correctly colored by our rules and this means that any diagram obtained from the trefoil by Reidemeister moves will inherit a coloring from this coloring that still has all three colors. (Think about this and you will see that it is so!). But this means that the trefoil can not be unknotted. For if it could then we would have transformed it to the unknot, and the unknot can only be colored with one color. So we have proved that the trefoil knot is knotted by using coloring.

Not every knot can be three-colored. For example, the figure eight knot cannot be so colored as the diagram below demonstrates. We start the coloring with two distinct colors a and b, propagate a c. Then the c interacts with a b and produces an a on a line already labeled with b. This contradiction shows that the figure eight knot cannot be three colored. This means that we have just proved that the figure eight knot is not isotopic to the trefoil knot, but we shall have to work harder to prove that the figure eight knot is actually knotted!

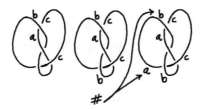

This can be done by using five colors and a more complex quandle but that is a story for another time.

I will end with one more kind of conclusion that we can draw from uncolorability. Consider the famous Borommean Rings as shown below. They are a link of three components. If you remove one of the rings, the other two come apart. We want to prove that the three rings cannot come apart. To do this, I give you an exercise. Prove that the Borommean Rings cannot be colored with three colors! You can verify this in a fashion similar to what we did with the figure eight knot. Now I will assume that you did this exercise and you are convinced that there is no way to color the rings.

But if the rings could come apart, then there would be a sequence of Reidemeister moves from the Borommean Rings to three unlinked rings. You can color each one of three unlinked rings with one of three different colors. The moves that got you from the Borormmean rings to the unlinked rings could be reversed and you would have a sequence of Reidemeister moves from the three unlinked rings to the Borommean rings. Each move would result in a three colored link,starting from the three unlinked colored rings. So in the end you would have to find a three coloring of the Borommean rings. That is a contradiction. Therefore the Borommean rings are linked. Is this not an amazing argument? (Nanyes 1993; Adams 1994).

Algebra and diagrams and their mathematical interpretaions interact in a multitude of ways that give rise to new ways to think about geometry, topology, algebra, combinatorics and indeed the entire mathematical universe.

The Roots of Lie Algebra

And now we return to the form of the epsilon. Let be given a trivalent vertex with sign change under permutation as we have had it from the beginning.

$$\curlyvee = -\curlyvee$$

And following our penchant to look at algebra in relation to diagrams let there be an algebra **L** so that the product of elements of **L** is indicated by the vertex.

Now contemplate a singular vertex as show below. In this singular vertex two arcs meet at a singular point along the arrow base-line.

There are three natural resolutions of this singularity and we have put them into a diagrammatic equation below.

This way to put the resolutions of this singularity into an equational pattern tells a nice algebraic story. In the algebra story we see that the equation is

$$(ab)c - (ac)b = a(bc)$$

and that this can be changed by using $b(ac) = -(ac)b$ to

$$(ab)c + b(ac) = a(bc).$$

This is called the *Jacobi Identity*.

An algebra that satisfies the Jacobi Identity and the anticommutativity of ab = − ba for all a and b in the algebra, is called a *Lie Algebra*. Lie algebras (Kauffman 2012; Bourbaki 1989) are ubiquitous in mathematics and indeed very closely related to the original epsilon of our paper and with the quandles in the knot theory, and more generally to knot theory in fundamental ways. It is quite surprising to meet the Jacobi identity as an expression of the resolution of a simple graphical singularity. Another story relates this combinatorics with the Reidemeister Moves (Kauffman 2012), but we will tell that tale another time.

But we cannot resist ending where we began and recount a little more of that conversation between RosePen and myself at the bar in the Ritz-Carlton. RosePen says to me: Are you familiar with Lie algebras? And I say, only a little. I know that the vector cross products form a Lie algebra and they satisfy the Jacobi Identity:

$$a \times (b \times c) = (a \times b) \times c + b \times (a \times c).$$

Well. He says. You can verify that Jacobi Identity by using the epsilon identity. I would not want to spoil the fun of it for you. Do it when you get back to your hotel room and before the rope tricks start this evening. I did, and I am sure the reader would like to do this as well. Once this exercise is completed the reader will see clearly that, enticing as it is, the epsilon is just the tip of the iceberg of a pattern to continues into Lie algebras, group theory, symmetry and beyond.

Epilogue

Some references may be useful to the reader. Much of the material in this paper can be found in the author's book "Knots and Physics" (Kauffman 2012) and in his papers (Kauffman 1990, 1992, 2005, 2016). The origin of the diagrammatics of vectors can be found in the work of Roger Penrose (Penrose 1971) and certain key insights and their diagrams are in the work of G. Spencer-Brown (Spencer-Brown 1979, 1997). For the coloring problem the reader can consult (Appel and Haken 1977; Kauffman 1990, 2016; Penrose 1971; Spencer-Brown 1979; Heawood 1890). For Lie algebras, a good start is (Stillwell 2008).

In this paper Professor RosePen is a fictional character who takes on some of the ideas and mathematical attitudes of Roger Penrose, George Spencer-Brown, John Horton Conway and the Author.

I have included some of my favorite mathematical tricks in this paper. The intent however is to go beyond tricks and ask about the nature of the sort of relationships that we have seen here. There are many more relationships of this kind. My field, topology, is full of them, and I am sure that other mathematicians in other fields would have many examples of their own. All of these examples use a diagram or some geometry to pivot between one conceptual domain and another. These diagrams give us an excuse to shift from one point of view to another and to find that the two points of view are related by the structure of the diagram and the meanings that are associated with it.

One can think about this situation as an allegory for a search for relationship that is mediated by a special place where the meeting can be accomplished. That place, the place of the diagram, is a multiplicity that is a unity where the multiplicity resides in the many interpretations that the diagram can receive, and the unity resides in the act of making the diagram, a making that can be accomplished and reenacted by any one who wishes to come to the understanding that the diagram offers. It is in the making that the many becomes the one and the one becomes the many.

References

K. I. Appel, W. Haken, Every planar map is four colorable. Part I: discharging, Illinois J. Math. 21 (1977), 429-490.

K. I. Apprl, W. Haken, J. Koch, Every planar map is four colorable. Part II: reducibility, Illinois J. Math. 21 (1977), 491-567.

L. H. Kauffman, A state calculus for graph coloring, Illinois Journal of Mathematics, Vol. 60, No. 1, (2016. pp. 251-271.

L. H. Kauffman, Map coloring and the vector cross product, J. Combin. Theory Ser. B 48 (1990), no. 2, 145–154.

L.H. Kauffman, Map coloring, q-deformed spin networks, and Turaev-Viro invariants for 3-manifolds, Internat. J. Modern Phys. B 6 (1992), no. 11–12, 1765–1794.

L. H. Kauffman, Reformulating the map color theorem, Discrete Math. 302 (2005), no. 1–3, 145–172.

L. H. Kauffman, Knots and Physics, 4th ed., World Scientific, Singapore, 2012.

A.B. Kempe, On the geographical problem of the fourc olors, Amer.J.Math.2 (1879), 193–201. MR 1505218

R. Penrose, Applications of negative dimensional tensors, Combinatorial mathematics and its applications (D. J. A. Welsh, ed.), Academic, San Diego, 1971.

G. Spencer-Brown, Laws of form [Gesetze der Form], Bohmeier Verlag, Leipzig, 1997. MR 0416823

G. Spencer-Brown, "Cast and Formation Properties of Maps," 1979. (unpublished)

P. J. Heawood, Map-colour theorem, Q. J. Math. Oxford 24 (1890), 332–338.

S.L. Altmann, Rotations, Quaternions, and Double Groups. Oxford University Press, Oxford, 1986.

Bourbaki, Nicolas (1989). *Lie Groups and Lie Algebras,* Springer-Verlag Pub.

Nanyes, O. An elementary proof that the Borommean rings are non-splittable. American Mathematical Monthly, 100(8):786–789, 1993.

Adams, C. *The Knot Book.* W. H. Freeman and Company, New York, 1994.

Stillwell, John. *Naïve Lie Theory*, (2008) Springer-Verlag, Pub.

Chapter 8
Experimental Mathematics: Overview and Pedagogical Implications

Marcel Danesi

Introduction

In 2011, the TV quiz show *Jeopardy* featured two human champions competing against IBM's Watson, an AI system designed for the event. Watson won the match by a large margin. Then, in 2017, another AI system, AphaGo, beat the world's Go champion with a creative move that was previously unknown, surprising Go experts. Popularized events such as these have made it saliently obvious to a large audience that AI bears many implications for understanding what intelligence is; and given that those AI systems were the product of a partnership between mathematicians and computer scientists, it is also obvious that they bear specific implications for how mathematics itself is practiced in the current technological environment, called the Information Age or, equally, the Computer Age. If an AI system can be devised to come up with a truly intelligent move in the game of Go, previously unbeknownst to humans, then the question arises: Can AI do creative mathematics? A positive answer does not seem to be beyond the realm of possibility.

The use of AI to solve mathematical problems and prove theorems has become its own discipline, called Experimental Mathematics (EM). As research in AI becomes ever more sophisticated, it might even be possible for a mechanical "super intelligence" to emerge, as Ray Kurzweil calls it in his 2005 book, *The singularity is near*, in which he maintains that there will come a moment in time when AI will have progressed to a point that it will autonomously outperform human intelligence. That moment, known as the *(technological) singularity*, will occur when an upgradable software becomes self-sufficient without human intervention, thus becoming capable of self-improvements. Each new self-improvement will bring about an intelligence explosion that will, in turn, lead to a powerful artificial

M. Danesi (✉)
Department of Anthropology, Fields Institute for Research in Mathematical Sciences, University of Toronto, Toronto, ON, Canada
e-mail: marcel.danesi@utoronto.ca

© Springer Nature Switzerland AG 2020

S. A. Costa et al. (eds.), *Mathematics (Education) in the Information Age*, Mathematics in Mind, https://doi.org/10.1007/978-3-030-59177-9_8

super-intelligence that will surpass human intelligence. Kurzweil predicts that the singularity should occur in the 2040s, when AI technologies will be so advanced that they cannot be controlled any longer by human intervention (Kurzweil 2012). By then, networks of silicon neurons will possess the same kind of information-processing functions of brain cells and thus operate at the speed of neurons.

It is relevant to note that the idea of a singularity can be traced back to a comment made by mathematician John van Neumann, cited by Stanislas Ulam (1958: 5): "[The] ever accelerating progress of technology and changes in the mode of human life, which gives the appearance of approaching some essential singularity in the history of the race beyond which human affairs, as we know them, could not continue." Then, in a 1965 essay, mathematician I. J. Good predicted that eventually an ultraintelligent machine would trigger an intelligence explosion (Good 1965: 31):

> Let an ultraintelligent machine be defined as a machine that can far surpass all the intellectual activities of any man however clever. Since the design of machines is one of these intellectual activities, an ultraintelligent machine could design even better machines; there would then unquestionably be an 'intelligence explosion,' and the intelligence of man would be left far behind. Thus the first ultraintelligent machine is the last invention that man need ever make.

In 1981, writer Vernor Vinge popularized Good's ideas, using the term *singularity* in his novella *True names*. He followed this up with a 1993 article in which he maintained that the singularity would become a reality in the first part of the twenty-first century. Whether the singularity is a realistic notion or not, the point is that EM fits into this scientific Zeitgeist, based on a mathematics-computer science partnership, raising several key questions regarding both mathematics and mathematics pedagogy. The purpose of this chapter is to look at these questions in a general way. If there is indeed a real possibility that AI can do mathematics independently of humans, then what does it tell us about mathematics? Can it discover new mathematics? What are the implications for mathematics education, if any?

Experimental Mathematics

A primary objective of EM is, literally, to carry out "experiments" with algorithms to see what they yield mathematically. These include creating computer programs with the ability to solve mathematical problems, prove theorems, and unravel patterns in classic conjectures. Although its origins go back to computer experiments on theorem proving in the 1950s, the first true texts in this field are the ones by Donald Knuth, which go back to the late 1960s. Knuth called the mathematics-computer science partnership *concrete mathematics* (see Graham et al. 1989), defining it simply as the analysis of algorithms, and the insights into the nature of mathematics that this provides. EM has achieved a number of impressive results, from discovering a formula for the binary digits of π (Bailey et al. 1997) to finding the smallest counterexample to the sum of powers conjecture by Euler (Frye 1988). As Borwein and Bailey (2004: vii) aptly point out, the importance of EM inheres in providing a concrete understanding of mathematical properties, by confirming or confronting conjectures, thus making mathematics "more tangible, lively and fun for both the professional researcher and the novice." EM has its own journal, *Experimental Mathematics*, founded in 1992, acknowledging its importance as a general research field in mathematics.

One of the primary forms of algorithmic experimentation involves machine learning, a term put forth by Arthur Samuel in 1959. This is concerned with designing computer programs that are capable of learning inductively. A famous feat of this kind of program came in 1996 when IBM developed a chess algorithm named Deep Blue that was capable of analyzing millions of chess positions every second and to learn from the evolving configurations of the chess pieces on the board, adjusting its program accordingly. Although it lost its first competition to chess champion Garry Kasparov, in a rematch it defeated him soundly. A subbranch of machine learning theory is now called Deep Learning AI; one of its aims is to study how computers can learn from huge amounts of data via artificial neural networks, which are computer networks that mimic the architecture and functions of neurons in the brain.

The first practical outcome of the mathematics-computer science alliance came in the mid-1950s with automated theorem proving (ATP) (Urban and Vyskočil 2013). ATP was at first based on the propositional calculus and predicate logic elaborated by Gottlob Frege (1879) and formalized later by Russell and Whitehead in their *Principia mathematica* (1913). It was called a first-order logical system. Russell and Whitehead thought they could derive all mathematics using axioms and the inference rules of formal logic, laying the foundations for ATP, even though this turned out to be overly optimistic (see below). The first ATP system, called a first-order proof system, was developed by Martin Davis in 1954. It was capable of solving a small set of logical theorems and grasping elementary mathematical properties. As Davis quipped about this early system (cited in Davis 2001: 3): "Its great triumph was to prove that the sum of two even numbers is even." In 1956, Newell, Simon and Shaw developed a Logic Theory Machine, based on the *Principia mathematica*, which had the ability to elaborate a small set of proofs based on the Russell-Whitehead logical system (see Newell and Simon 1956). With human guidance, the system was able to prove 38 of the first 52 theorems of the *Principia*. This approach was called "heuristic" because the Machine attempted to mimic human mathematicians, but it could not ensure that a proof could be carried out for every valid theorem. Since then, more sophisticated algorithms have been designed which, in theory, could prove or assess any theorem based on first-order logic. But, despite such sophistication, ATPs at present are mainly capable of solving elementary mathematical problems and carrying out basic proofs. As Ganesalingam and Gowers (2017: 253) point out, devising a truly intelligent ATP system is a highly realizable goal, but is still in its fledgling stages:

> The main challenge…is that it does not seem to be possible to reconstruct genuinely human-like writeups from the proofs produced by automated provers from the machine-oriented tradition. In order to be able to produce such writeups we have had to use much more restricted proof methods than those available to modern provers. This in turn makes it

more challenging for the prover to solve any individual problem, and indeed the program does not solve any problems that are beyond the reach of existing fully automatic provers. We should also note that we see this prover as the first stage of a long-term project to write a program that solves more complex problems in a 'fully human' way.

ATP has already helped prove a number of hard theorems, such as the Robbins conjecture, which had eluded human mathematicians, until William McCune devised a computer program that proved it (McCune 1997). The event was one of the first to raise the possibility that computers could actually do mathematics, perhaps even better than humans. McCune called his program EQP, for *equational prover*. He checked the proof by computer and by hand. It was then checked independently by other mathematicians. To distance himself from the ATPs of the past, McCune called his project an experiment in "automated reasoning," rather than "automated theorem proving." It was, arguably, one of the first "ultraintelligent" algorithms, to use Good's (1965) designation (above).

The Entscheidungsproblem

For "automated mathematics" to be able to do mathematics, it would require a system of axioms and rules that could be applied to all problems. In other words, for AI to be able to do mathematics, independently of humans, the mathematics fed into it would have to be consistent and able to decide if a problem is solvable or not.

With the *Principia mathematica*, Russell and Whitehead aimed to provide such a system. As they explain in their introduction, their aim was (Russell and Whitehead 1913: 1): (1) to analyze the ideas and methods of mathematical logic, minimizing the number of primitive notions and axioms, and inference rules; (2) to precisely express mathematical propositions in symbolic logic using the most efficient notation possible; and (3) to solve the paradoxes that plagued logic and set theory, such as the Liar Paradox of antiquity and Russell's own Barber Paradox, which involved circularity in thinking.

But the *Principia* did not overcome the *Entscheidungsproblem* ("decision problem"), which originated with Leibniz, who wanted to build a machine that could determine the truth values of mathematical statements, realizing that this would entail developing a consistent formal language (Davis 2000: 3–20). The problem was elaborated explicitly by David Hilbert and Wilhelm Ackermann in 1928. Hilbert (1928) believed that there was no such thing as an undecidable or unsolvable problem. But Kurt Gödel's (1931) famous incompleteness theorem that within any formal set of propositions, such as those in the *Principia,* there are some that can be neither proved nor disproved, showed that mathematical systems are de facto incomplete. For the present purposes, the main argument put forth by Gödel can be condensed as follows (from Danesi 2002: 146):

> Consider a mathematical system that is both correct—in the sense that no false statement is provable in it—and contains a statement "S" that asserts its own unprovability in the system. S can be formulated simply as: "I am not provable in system T." What is the truth status

of S? If it is false, then its opposite is true, which means that S is provable in system T. But this goes contrary to our assumption that no false statement is provable in the system. Therefore, we conclude that S must be true, from which it follows that S is unprovable in T, as S asserts. Thus, either way, S is true, but not provable in the system.

A few years later, Alonzo Church (1936) demonstrated why a solution to the Entscheidungsproblem is unlikely, employing a method that would now fall within the EM paradigm. According to Church, the Entscheidungsproblem could be approach meaningfully only if the notion of algorithm was formally defined. In 1936 Alan Turing had also examined the Entscheidungsproblem in terms of his "Turing machine"—a mathematical model of a hypothetical computing machine which can use a set of predefined rules to determine a result from a set of input variables. Church proved that there is no computable function which decides if two given expressions are equivalent or not. For Turing, the Entscheidungsproblem can similarly be framed in computational terms as devising an algorithm capable of deciding whether a given statement is provable from the axioms using the rules of first-order logic. Turing showed that this was impossible, calling it the Halting Problem. Given a computer program and an input, will the program finish running or will go into a loop and run forever? Turing argued that no algorithm for solving this problem can exist logically. Another formulation of the Halting Problem is as follows: Can a computer program be envisioned that can look at any other program and decide if it will ever stop running? Here is a paraphrase of Turing's proof by contradiction:

> Assume that there is such a program. If so, we could run it on a version of itself, which would halt if it determines that the other program never stops, and runs an infinite loop if it determines that the other program stops. This is a contradiction.

The Church-Turing thesis, as it came to be known, presented an early obstacle to AI systems such as ATPs, since they were subject to incompleteness and to the Halting Problem—problems that remain unsolved to this day. Perhaps an ultraintelligent machine—to recycle that phrase—can overcome these two obstacles to true AI, both of which derive from the nature of the human brain. Of course, this is speculation, but nonetheless an interesting form of speculation.

The P = NP Problem

Another problem facing automated mathematics is the P = NP problem. It asks if every problem whose solution can be quickly checked or verified can also be solved quickly. This can be illustrated simply with the game of Sudoku (Fortnow 2013). Any proposed solution to the game can be easily realized by a computer algorithm; but the time it takes to check the solution grows slowly (that is, polynomially) as the grid gets larger. Sudoku is thus said to be in P (quickly solvable), but not in NP (quickly checkable). Thousands of other problems are similar; that is, they can be realized quickly, but take longer and longer times to validate as they increase in complexity. Polynomial time refers to the time a computer takes as a polynomial function of the size of the input. The expression "quickly solvable" means in this framework that an algorithm can solve a task in polynomial time—that is, it is a polynomial function of the size of the input to the algorithm. The general class of problems for which an algorithm can provide an answer in polynomial time is called P, which stands for "polynomial time." Another class of problems are those for which there is no quickly solvable answer, but which can nonetheless be verified in polynomial time if relevant information is inputted. This class of problems, for which an answer can be verified in polynomial time, is called NP, which stands for "nondeterministic polynomial time."

Computer scientists have shown that problems in NP have the property that a fast solution to any one of them can be used to build a quick solution to any other problem, a property called NP-completeness. If P were equal to NP then problems that are complex (involving large amounts of data) could be tackled easily as the algorithms become more efficient. The P = NP problem is one of the most important open problems in computer science and mathematics—it would determine whether problems that can be verified in polynomial time can also be solved in polynomial time. If it turned out that P ≠ NP, which is widely believed to be the case, it would mean that there are problems in NP that are harder to compute than to verify: that is, they could not be solved in polynomial time, but the answer could be verified in polynomial time.

The formal articulation of the P = NP problem is traced to a 1971 paper by Stephen Cook (and independently a few years later to another paper by Leonid Levin 1973). An early mention of the problem is found in a 1956 letter written by Kurt Gödel to John von Neumann. Gödel asked Neumann whether theorem-proving could be solved in quadratic or linear time—a central question for automated mathematics (Fortnow 2013). P would consist of all those problems that can be solved on a deterministic sequential machine in an amount of time that is polynomial in the size of the input; the class NP would consist of all those problems whose solutions can be verified in polynomial time given the relevant information, or equivalently, whose solution can be found in polynomial time on a non-deterministic machine. Some of the existing automated proof techniques are not powerful enough to answer the P = NP problem, suggesting that novel technical approaches are required. It is those approaches that occupy a great amount of interest within EM.

Autopoiesis

Another major obstacle to fully automated mathematics as a surrogate for human-based mathematics concerns the nature of human creativity itself. The latter involves so-called embodied cognition, an idea traced to the theory of *autopoiesis* (Maturana and Varela 1973). For the purposes of the present discussion, this implies that human thought organizes itself on the basis of the input received through the body; that is, by self-organizing input changes through specific anatomical and sensory-feeling systems. The latter were called the constituents of the human mental *Bauplan* by the biologist Jakob von Uexküll (1909) at the turn of the twentieth century. For von Uexküll, creativity emerges through the ability of an organism to organize itself creatively in responses to environmental changes. Animals with widely divergent brains (or neural systems) do not generate the same kinds of creative responses to input. There exists, therefore, no common world of objects shared by humans and non-human animals equally—and by extension, by humans and machines.

Can automated reasoning systems become autopoietic in the human sense? As Maturana and Varela (1973: 16) observe, autopoiesis "takes place in the dynamics of the autonomy proper to living systems." An example of an autopoietic system is the eukaryotic cell, which is organized into structures such as the nucleus, organelles, a membrane and cytoskeleton. These depend on an external flow of molecules and energy which, in turn, allow for these very components to organize themselves (like a wave propagating itself through a medium). Autopoietic systems are thus self-propagating and self-contained. These are contrasted to allopoietic systems, such as an automobile assembly line, which involves assembling raw materials into an automobile (an organized structure), which is something other than itself (the assembly line). Human creativity is autopoietic; automated reasoning, like an assembly line, is (currently) largely allopoietic. It is when the two are combined, as in the proof of the Robbins conjecture, that the results are truly remarkable. McGann (2000: 358) provides the following relevant characterization of autopoiesis:

> An autopoietic system is a closed topological space that continuously generates and specifies its own organization through its operation as a system of production of its own components, and does this in an endless turnover of components. Autopoietic systems are thus distinguished from allopoietic systems, which are Cartesian and which have as the product of their functioning something different from themselves. Coding and markup appear allopoietic.

Among the main allopoietic features of automated reasoning systems are the following:

1. They can describe input data correctly but cannot make any non-trivial predictions or hypotheses about the underlying system.
2. They are limited to technology and to the mathematics used, making follow-up experimental testing trivial, since there would be no known real world counterparts to the theorized model.
3. However, if the model results in a non-trivial or unexpected experimental hypothesis, by happenstance, then it can be tested and verified by further human intervention. This may lead to the design and implementation of new automated experiments and may lead, in turn, to potentially significant results.

As Lakoff and Núñez (2000) have shown, solving problems and doing mathematical proofs involves a feature of language that is maximally autopoietic—metaphorical reasoning. Lakoff and Núñez characterize many problems and proofs as being based on *grounding* and *linking* metaphors. The former encode basic ideas, being directly grounded in experience. For example, addition develops from the experience of counting objects and then inserting them in a collection. Linking metaphors connect concepts within mathematics that may or may not be based on physical experiences. Some examples of this are the number line, inequalities, and absolute value properties within an epsilon-delta proof of limit. Linking metaphors are the source of negative numbers, which also emerge from an autopoietic form of reasoning that, as Alexander (2012: 28) elaborates, is an "entity with its own identity:"

> Using the natural numbers, we made a much bigger set, way too big in fact. So we judiciously collapsed the bigger set down. In this way, we collapse down to our original set of natural numbers, but we also picked up a whole new set of numbers, which we call the negative numbers, along with arithmetic operations, addition, multiplication, subtraction. And there is our payoff. With negative numbers, subtraction is always possible. This is but one example, but in it we can see a larger, and quite important, issue of cognition. The larger set of numbers, positive and negative, is a cognitive blend in mathematics…The numbers, now enlarged to include negative numbers, become an entity with its own identity. The collapse in notation reflects this. One quickly abandons the (minuend, subtrahend) formulation, so that rather than (6, 8) one uses -2. This is an essential feature of a cognitive blend; something new has emerged.

Mathematics makes sense when it encodes concepts that fit our experiences of the world—experiences of quantity, space, motion, force, change, mass, shape, probability, self-regulating processes, and so on. The inspiration for new mathematics comes from these experiences as it does for new language. It is unclear how AI can ever *experience* input in this way. And if it did, what would it make of a truly creative blend? It would still take human intervention to *interpret* it. The bimodal morphology of the human brain, with its two hemispheres, is the root of creativity (Gardner 1982: 74). When the brain takes in unfamiliar information it requires the experiential (probing) right-hemisphere functions to operate freely to grasp it; these can be called R-Mode functions (Danesi 2017). However, this exploratory effort would be virtually wasted if not followed up by the brain's ability to simultaneously analyze the same input; this is a left-hemispheric capacity that can be called an L-Mode function. So, human creativity can be formulated as a blending of information between the R-Mode and the L-Mode. The creation and solution of some problem involves the R-Mode and L-Mode interacting dynamically. The work of Russian psychologist Lev S. Vygotsky (1961) is relevant to this bimodal view of the brain. Vygotsky's research on children has shown that creativity manifests itself in a blend of non-verbal symbolism (play, drawing, etc.), which has its source in the R-Mode, and verbal symbolism (narratives, fables, dramatizations, etc.), which has its source in the L-Mode. The emergence of creative, abstract thinking emerges as the two modes interact dynamically in even the simplest of tasks.

AI has taken great strides in advancing how we may indeed construct some L-Mode systems of representation. As far as can be told, R-Mode systems, such as those involved in metaphorical reasoning, are lacking (see Black 1962). As physicist Roger Penrose (1989) has argued, computers can never truly be creative in the human sense because the laws of nature will not allow it.

Prospects

Despite the obstacles mentioned above, Experimental Mathematics is a thriving field, with automated systems providing insights to human mathematicians that would literally have been unthinkable in the past. Neural network theory, for example, has been remarkably powerful in this sense. It was used to create the AI that beat the world champion at Go, mentioned at the start of this chapter. The program has been developed further to play against itself using reinforcement learning. This type of program, called DeepMind, is a very useful one mathematically. It receives as input a series of equations along with their solutions, without any explanation of how those solutions can be reached. In recent versions, DeepMind was devised to "intuit" how to solve equations without any instructions or structure, based solely on examining a limited number of completed examples. As opposed to other AI systems, DeepMind is not pre-programmed, learning from experience. Of course, the experience is based purely on pattern-extraction from data that is introduced into the system by humans. It is not fully autopoietic.

To study creativity, a subfield of AI has emerged, called computational creativity (McCormack and D'Inverno 2012). The problem is that the notion of human creativity is still an intuitive one, and impossible to define—a criterion that is essential to any computational approach. What would creativity be like in an AI system? Would it need to be autopoietic in the human sense? Among the first computer scientists to tackle the problem of computational creativity were Newell, Shaw, and Simon in 1963, who eliminated the wide-open meaning associated with creativity, defining it more narrowly as something that is novel and useful. A computer can indeed come up with something novel, as the case of the Go match has shown. But was it creative in the human sense? The computer examined a database of about 100,000 human Go matches, playing against itself millions of times, reprogramming and improving itself, using a Monte Carlo tree search algorithm based in neural network theory to carry out an "analysis" of a winning move.

Suffice it to say that the Newell-Shaw-Simon approach has had implications for AI generally. For example, Schmidhuber (2010) has argued that creativity is based on a simple computational principle for optimizing learning—the processing and encoding of a continually growing history of actions and inputs, implemented with an artificial neural network or some other machine learning device that can exploit patterns to improve performance over time. According to Schmidhuber, this explains all kinds of human creative acts, including discovering new mathematical ideas and theorems. So far, however, there is no evidence that human and computational creativity share the same principle. Recall that autopoiesis involves autonomous self-organization. So can computers really become autonomous in this sense? It is suggestive to note that Google has developed its own AI theorem-proving program, which can prove, essentially unaided by humans, many basic theorems of mathematics. In 2019, two members of Facebook's AI research group, Guillaume Lample and François Charton, even developed a neural network system capable of solving symbolic mathematical problems.

Despite such truly remarkable feats, automated mathematics remains an allopoietic intelligence at the present time. Experimental Mathematics is important as a kind of laboratory of allopoietic mathematics, where the laboratory is the computer, providing a concrete way for human mathematicians to gain insights, unravel mathematical principles, test conjectures, and confirm results.

Pedagogical Implications

The Robbins conjecture proof (above) laid the groundwork for what is now called the "computational thinking" method in mathematics pedagogy (Papert 1980; Wing 2006). The characteristics that define it are problem deconstruction, pattern recognition, data representation, abstraction, and algorithm-creation. Simply put, it envisions the learning of mathematics as being implanted on learning how to program computers to solve problems. Papert (1980) saw the goal of classroom pedagogy as creating a "programming environment" that invites students to write codes to solve specific problems. David Mumford and Sol Garfunkel (2013: 174) elaborate this pedagogical principle as follows:

> Everyone says computer technology should be used in schools, but why let the computer be another incomprehensible technological mystery? Teach everyone the rudiments of programming and what goes on inside that box. "But is this math?" we hear you saying. Yes; writing computer code teaches you how to be precise and formal and makes concrete mathematical recipes like that for long division. They are what we call algorithms, and this sort of training is a paradigm for rational thinking.

The central idea in this kind of pedagogy is that, by trying to figure out how to design algorithms to solve problems of various sorts, learners can discover mathematical patterns and even identify errors in their previous thinking. Kosslyn (1983: 116) described this aspect of computers aptly decades ago:

> The computer model serves the function of a note pad when one is doing arithmetic: It helps keep track of everything so that you don't get a headache trying to mentally juggle everything at once. Sometimes the predictions obtained in this way are surprising, which often points out an error in your thinking or an unexpected prediction.

AI systems can now respond to student questions about specific topics, identify what aspects the student is ready to learn, which ones need more training, and so on (Cheney et al. 2011; Nansen et al. 2012; McCoy 2014). There are also a number of computer-based mathematics projects for in-class pedagogy. One of the most comprehensive has been designed by Conrad Wolfram (2020), who believes that mathematics education should be based entirely on developments in the AI mathematics laboratory.

Does this mean that the teacher has no role to play any longer? Like the need for human mathematicians to guide automated mathematics, so too there is a need for human teachers to guide the whole process. Indeed, paradoxically, computer-based learning puts even more of an onus on the teacher, as McLuhan and Leonard (1967: 24) remarked decades ago:

> Tomorrow's educator will be able to set about the exciting task of creating a new kind of learning environment. Students will rove freely through this place of learning, be it contained in a room, a building, a cluster of buildings or an even larger schoolhouse. There will be no distinction between work and play in the new school, for the student will be totally involved. Responsibility for the effectiveness of learning will be shifted from student to teacher.

Concluding Remarks

The question of whether AI can do mathematics is ultimately a rhetorical one. When mathematicians talk about something coming into existence as the result of some proof or serendipitous discovery, they are talking about something that only humans can truly interpret and understand. As Ian Stewart (2013: 313) observes, the problem of *existence* is hardly a trivial one:

> The deep question here is the meaning of "exist" in mathematics. In the real world, something exists if you can observe it, or, failing that, infer its necessary presence from things that can be observed. We know that gravity exists because we can observe its effects, even though no one can see gravity. However, the number two is not like that. It is not a thing, but a conceptual construct.

The irrational numbers and the imaginary ones did not "exist" until they cropped up in the solution of two specific equations made possible by the Pythagorean theorem and the concept of quadratic equation respectively. So, where were they before? Were they waiting to be discovered? This question is clearly at the core of the nature of mathematics. These did not "exist" until they crystallized in the conduct of mathematics, through ingenious notational modifications, diagrammatic insights, ludic explorations with mathematical signs, and so on. These features of the human mind are likely to be beyond the grasp of automated systems—at least at the present time.

References

Alexander, J. (2012). On the cognitive and semiotic structure of mathematics. In: M. Bockarova, M. Danesi, and R. Núñez (eds.), *Semiotic and cognitive science essays on the nature of mathematics*, 1–34. Munich: Lincom Europa.

Bailey, D. H., Borwein, J. M., and Plouffe, S. (1997). The quest for Pi. *Mathematical Intelligencer* 19: 50–57.

Black, M. (1962). *Models and metaphors*. Ithaca: Cornell University Press.

Borwein, J. M. and Bailey, D. (2004). *Mathematics by experiment: Plausible reasoning in the 21st century*. Natick, Mass.: A. K. Peters.

Cheney, K. R., Craig, S. D., Anderson, C., Bargagliotti, A., Graesser, A. C., Sterbinsky, A., Okwumabua, T. and Hu, X. (2011), Closing the knowledge gap in mathematics among sixth grade students using ALEKS. In: M. Koehler and P. Mishra (eds.), *Proceedings of Society for Information Technology & Teacher Education International Conference 2011*, 1425–1427. Chesapeake, VA: AACE.

Church, A. (1936). An unsolvable problem of elementary number theory. *American Journal of Mathematics* 58: 345–363.

Cook, S. (1971). The complexity of theorem proving procedures. *Proceedings of the Third Annual ACM Symposium on Theory of Computing*, 151–158. https://doi.org/10.1145/800157.805047.

Danesi, M. (2002). *The puzzle instinct: The meaning of puzzles in human life*. Bloomington: Indiana University Press.

Danesi, M. (2017). *Puzzles in mathematics education*. München: Lincom Europa.

Davis, M. (2000). *Engines of logic*. London: W.W. Norton.

Davis, M. (2001). The early history of automated deduction. In: A. Robinson and A. Voronkov (eds.), *Handbook of automated reasoning*, 3–15. Oxford: Elsevier.

Fortnow, L. (2013). *The golden ticket: P, NP, and the search for the impossible*. Princeton: Princeton University Press.

Frege, G. (1879). *Begiffsschrift eine der Aritmetischen nachgebildete Formelsprache des reinen Denkens*. Halle: Nebert.

Frye, R. E. (1988). Finding $958004 + 2175194 + 4145604 = 4224814$ on the connection machine. *Proceedings of supercomputing 88*, 106–116. Silver Spring: IEEE Computer Society.

Ganesalingam, M. and Gowers, W. T. (2017). A fully automatic theorem prover with human-style output. *Journal of Automated Reasoning* vol 58: 253–291.

Gardner, H. (1982). *Art, mind, and brain: A cognitive approach to creativity*. New York: Basic.

Gödel, K. (1931). Über formal unentscheidbare Sätze der Principia Mathematica und verwandter Systeme, Teil I. *Monatshefte für Mathematik und Physik* 38: 173–189.

Good, I. J. (1965). Speculations concerning the first ultraintelligent machine. *Advances in Computers* 6: 31–83.

Graham, R., Knuth, D., and Patashnik, O. (1989). *Concrete mathematics: A foundation for computer science*. New York: Addison-Wesley

Hilbert, D. (1928). Die Grundlagen der Mathematik. *Abhandlungen aus dem Seminar der Hamburgischen Universität* 6: 65–85.

Hilbert, D. and Ackermann, W. (1928) *Grundzüge der theoretischen Logik*. Berlin: Springer.

Kosslyn, S. M. (1983). *Ghosts in the mind's machine: Creating and using images in the brain*. New York: W. W. Norton.

Kurzweil, R. (2005). *The singularity is near*. Harmondsworth: Penguin.

Kurzweil, R. (2012). *How to create a mind: The secret of human thought revealed*. New York: Viking.

Lakoff, G. and Núñez, R. (2000). *Where mathematics comes from: How the embodied mind brings mathematics into being*. New York: Basic Books.

Lample, G. and Charton, F. (2019). Deep learning for symbolic mathematics. arXiv:1912.01412.

Levin, L. A. (1973). Universal search problems. *Problems of Information Transmission* 9: 115–116.

Maturana, H. and Varela, F. (1973). *Autopoiesis and cognition*. Dordrecht: Reidel.

McCormack, J. and d'Inverno, M. (eds.) (2012). *Computers and creativity*. Berlin: Springer.

McLuhan, M. and Leonard, G. B. (1967). The future of education: The Class of 1989. *Look*, February 21, pp. 23–24.

McCoy, L. (2014). Computer-based mathematics learning. *Journal of Research on Computing in Education* 28: 438–460.

McCune, W. (1997). Solution of the Robbins problem. *Journal of Automated Reasoning* 19: 277–318.

McGann J. (2000). Marking texts of many dimensions. In: S. Schreibman, R. G. Siemens, and J. M. Unsworth (eds.), *A companion to digital humanities*, 358–376. Hoboken: John Wiley & Sons.

Mumford, D. and Garfunkel, S. (2013). Bottom line on mathematics education. In: M. Pitici (ed.), *The best writing in mathematics 2012*, 173–175. Princeton: Princeton University Press.

Nansen, B., Chakraborty, K., Gibbs, L., Vetere, F., and MacDougall, C. (2012). You do the math: Mathletics and the play of online learning. *New Media & Society* 14: 1216–1235.

Newell, A., Shaw, J. G., and Simon, H. A. (1963). The process of creative thinking. In: H. E. Gruber, G. Terrell and M. Wertheimer (eds.), *Contemporary approaches to creative thinking*, 63–119. New York: Atherton.

Newell, A. and Simon, H. A. (1956). *The Logic Theory Machine. A complex information processing system*. Santa Monica: The Rand Corporation.

Papert, S. (1980). *Mindstorms: Children, computers, and powerful ideas*. New York: Basic Books.

Penrose, R. (1989). *The emperor's new mind*. Cambridge: Cambridge University Press.

Russell, B. and Whitehead, A. N. (1913). *Principia mathematica*. Cambridge: Cambridge University Press.

Samuel, A. (1959). Some studies in machine learning using the game of checkers. *IBM Journal of Research and Development* 3: 210–229.

Schmidhuber, J. (2010). Formal theory of creativity, fun, and intrinsic motivation. *IEEE Transactions on Autonomous Mental Development* 2: 230–247.

Stewart, I. (2013). *Visions of infinity*. New York: Basic Books.

Turing, A. (1936). On computable numbers with an application to the Entscheidungsproblem. *Proceedings of the London Mathematical Society* 41: 230–265.

Uexküll, J. von (1909). *Umwelt und Innenwelt der Tierre*. Berlin: Springer.

Ulam, S. (1958). Tribute to John von Neumann. *Bulletin of the American Mathematical Society* 64: 5.

Urban, J. and Vyskočil, J. (2013). Theorem proving in large formal mathematics as an emerging AI field. In: M. P. Bonacina and M. E. Stickel M.E. (eds), *Automated reasoning and mathematics*. Berlin: Springer.

Vinge, V. (1981). *True names*. New York: Tor Books.

Vinge, V. (1993). The coming technological singularity: How to survive in the post-human era. In: G. A. Landis (ed.), *Vision-21: Interdisciplinary science and engineering in the era of cyberspace*, 11–22. NASA Publication CP-10129.

Vygotsky, L. S. (1961). *Thought and language*. Cambridge, Mass.: MIT Press.

Wing, J. M. (2006). Computational thinking. *Communications of the ACM* 49 (3): 33. doi:https://doi.org/10.1145/1118178.1118215.

Wolfram, C. (2020). *The math(s) fix: An education blueprint for the AI age*. Wolfram Media.

Chapter 9
Why the Basics Still Matter: The Cost of Using a Machine to Do Mathematics

Sasha Gollish

Introduction

Remember the joke 1 + 1 = a window? If you rearrange the numbers and symbols you can draw a simple window. It is a playful answer, and quite creative. This creativity based on playfulness is commonly missing these days from mathematics courses. Dan Finkel (2019), founder of For the Love of Math, says that when a student comes to a teacher with a playful solution such as the one above the response should always be "yes," because "yes," compared to "correct," starts a true dialogue between student and teacher, because "yes" says "I value and accept your idea;" "yes" is a mark of respect, and "no" can be a motivation destroyer.

I liked math because it was correct or not correct and I could fumble around to find the correct solution. I liked it because finding the right answer gave me confidence. I liked math because a teacher could not tell me the answer was wrong when I thought it was right, which was how I felt about reading comprehension and the purpose of a passage—at least now I can let books invoke feelings in me and no one tells me I'm wrong!

Developing the fluidity to move through the basic math operations gave me confidence when it came to do more complex math. I remember in OAC (Ontario Academic Credit) calculus that my peers would convert fractions to decimals, experiencing the whole process as a nightmare. Looking back, I can see that they missed out on the simplicity and beauty of seeing connections between fractions and decimals, because they were "afraid" of fractions. Plus, they needed a calculator to get to their decimals and what was usually a two-digit number (one on the top of the fraction, the other on the bottom) became a long string of numbers that was so much more cumbersome to work with.

S. Gollish (✉)
University of Toronto, Toronto, Canada
e-mail: s.gollish@utoronto.ca

© Springer Nature Switzerland AG 2020
S. A. Costa et al. (eds.), *Mathematics (Education) in the Information Age*,
Mathematics in Mind, https://doi.org/10.1007/978-3-030-59177-9_9

I understand why using a calculator was helpful to them. I will admit that I too will occasionally break down and use calculator on my phone to help me with something simple, but first I usually try to work it out in my head. That threshold for knowing the basics might be slipping, not just in mathematics, but all subjects. In a way, we all walk around with a device (i.e. any smart phone connected to a cellular data network) in our pocket, purse, bag, etc. that makes us a polymath. But relying on our machines does not really mean that we have "learned" that information, nor does it make us a polymath—a smart phone is simply a tool that supports knowledge.

Stepping back from the kind of technological world in which we live and thinking back to mathematics, the idea of knowing the fundamental (rudimentary) operations for addition, subtraction, multiplication, and division remains a crucial skill that all of us need, but equally as important is the ability to have fun with mathematics. There are so many fun things in life where we can use mathematics, from calculating a mortgage payment on your first home, to grocery shopping when you get your first adult job paycheck and you splurge on expensive cheese, even understanding geometry for parking your car in a spot at the grocery store to get that elusive "close" spot.

I have this memory of watching a British Minister take a phone out of her pocket and tell the world this was why we no longer needed to teach math. I may have imagined this event, or I may have a biased recollection to fit a narrative I want to tell myself. Regardless, I could not disagree more. I know that math gives some people the shivers, invokes horrible memories, and may even induce panic attacks. Unfortunately, I think that this is brought about because people may not have mastered the basics of math, which disadvantages them in many ways, and so they miss out on some very important things in life that involve knowing mathematics. This chapter will argue that we still need to teach students the basics and not rely on our machines to do mathematics.

What Are the Basics of Mathematics?

It is important to frame what is meant by basics of mathematics for the purposes of this chapter. The basic or foundational skills vary by age and ability. When one is first introduced to mathematics, usually as a child, it involves intuitive pattern recognition and counting. That early stages form a foundation for mathematics. As one progresses, intuitive forms become increasingly more formalized, turning into knowledge of simple arithmetic and subtraction, and then multiplication and division. Some of us might remember the 12 by 12 blocks we had to fill in, usually for addition and multiplication, as well as practicing multiplication and division with "mad minute" flash cards, or other heuristics devices.

Fields that require advanced mathematics are many, including doctors, engineers, scientists, etc., involving the calculus, algorithmic thinking, or estimation, which simply extend the basic mathematics skills (Gollish 2019). Defining the basics for mathematics has to take into account how math is employed, from early learning (number counting, pattern recognition), primary education (addition, multiplication, division, and subtraction), to advanced skill-acquisition (medicine and engineering). In effect, all such skills are based on grasping the foundational principles of mathematics that manifest themselves in our everyday lives, as from understanding a mortgage to how much one can spend on groceries per week or how many kilometers someone might need to run to train for a marathon.

In the Merriam-Webster dictionary, "fundamental" is defined as "serving as a basis supporting existence, determining essential structure; relating to essential structure, function, or facts; of central importance; and one of the minimum constituents without which a thing or a system would not be what it is" (Merriam-Webster 2020). "Basic" is a synonym of fundamental and is defined as "constituting or serving as the basis or starting point or concerned with fundamental scientific principles" (Merriam-Webster 2020).

It is the common set of basic mathematics skills that make one numerically literate, or numerate. Numeracy "is the ability to access, use, interpret, and communicate mathematical information and ideas, to engage in and manage mathematical demands of a range of situations in adult life" (National Centre for Education Statistics (2020). The Organization for Economic Cooperation and Development (OECD) builds upon this definition and adds that mathematical literacy "includes concepts, procedures, facts, and tools to describe, explain, and predict phenomena. It helps individuals know the role that mathematics plays in the world and make the well-founded judgments and decisions needed by constructive, engaged and reflective 21st Century citizens" (OECD 2019). The OECD is an umbrella organization for the Programme of International Student Assessment (PISA) which holds mathematics contests, and which serves as a way to rank countries in their mathematics skills. The PISA 2021 Mathematics Framework identified the following key aspects of foundational or basic mathematical reasoning:

- understanding quantity, number systems and their algebraic properties;
- appreciating the power of abstraction and symbolic representation;
- identifying mathematical structures and their regularities;
- recognizing functional relationships between quantities;
- using mathematical modelling as ways to grasp the real world (as in the physical, biological, social, economic and behavioural sciences); and
- understanding variation as the function of statistics.

These key traits are what one needs to be numerate. PISA emphasizes that basic mathematical cognition involves an ability to reason logically and present arguments so that one can "formulate, employ, and interpret and evaluate to solve problems in a variety of real-world contexts" (OECD 2019). This definition demonstrates that learning math involves a set of standards but competencies and outcomes that every individual should acquire in school. In Canada, provinces set their own standards or outcomes for mathematics education with the objective of imparting numerical literacy and reasoning (Council of Ministers of Education, Canada 2020). Other than Ontario and Quebec, provinces follow the Common Core Standards for Mathematics from the National Council of Teachers of Mathematics (NTCM), an

American institution. The common core standards are modeled after the method of "process and proficiency," which includes the following (Council of Ministers of Education, Canada 2020):

1. Making sense of problems and persevering in solving them.
2. Reasoning abstractly and quantitatively.
3. Constructing viable arguments and critiquing the reasoning of others.
4. Modeling with mathematics.
5. Using appropriate math ideas strategically.
6. Attending to precision.
7. Looking for and making use of structure.
8. Looking for and expressing regularity in repeated reasoning.

In Ontario, basic mathematics is "fostered through instruction that highlights strategies for remembering facts, focuses on making sense, and integrates math-fact learning into other aspects of math learning, such as developing computational skills." The document goes on to suggest that this is not to be interpreted as repeated practice or "drills," although it is not clear what is meant by this (Ministry of Education of Ontario 2005). The idea seems to be that math fluency correlates with the cognitive load—the curriculum is based on the belief that there should be a gradual process of learning, so that students should master addition and subtraction facts by the end of grade three and multiplication and division facts by the end of grade five. If that is the desired outcome practice, repetition seems to be the quickest means to an end but does not ensure that students understand, or reflect, upon how these operations function.

From any curriculum program, it is confusing to determine what constitutes the basics or foundational skills of mathematics. Certainly, "counting, adding and sub-tracting, multiplying, dividing, calculating change, calculating tips, and percent-ages," as suggested by Vinner (2018) in *Mathematics, education, and other endangered species*, are obvious aspects of math competence. Saul Khan, founder of Khan Academy, a non-profit organization dedicated to providing free, world-class education, which first started with a series of mathematics videos, breaks down math competence in the early stages as follows (Khan Academy 2020):

1. Counting (up to 120)
2. Addition and Subtraction (making 10, then 20, then 100, finally 1000)
3. Place Value (tens, hundreds)
4. Measurement and Data
5. Geometry

These are not trivial, especially when one puts a lens on how we teach these to our students. At the root of it there is a desire to ensure that all students become numerically literate (numerate) and to hopefully develop some passion and desire to pursue and play with mathematics, as opposed to dreading it.

How We Teach Mathematics

To talk about how to teach the basics (or fundamentals) of mathematics, it is necessary to first define what mathematics is, albeit schematically. Generally speaking, mathematics is "the science of numbers and their operations, interrelations, combinations, generalizations, and abstractions; of space configurations and their structure, measurement, transformations, and generalizations; and algebra, arithmetic, calculus, geometry, and trigonometry are branches of *mathematics*" (Merriam-Webster 2017). One might say, as seventy-five prestigious Canadian and American mathematicians of 1962 eloquently said, "to know mathematics means to be able to do mathematics: to use mathematical language with some fluency, to do problems, to criticize arguments, to find proofs, and, what may be the most important activity, to recognize a mathematical concept in, or to extract it from, a given concrete situation" (Ahlfors et al. 1962).

Defining mathematics leads to several ambiguities, unlike other subjects that generally have consistent definitions. This is borne out by the diversity in answers when one asks students and instructors to explain what mathematics is (Boaler 2015). Mathematics education researcher Jo Boaler (2015) found the following: if you ask a student what the study of mathematics is the student will tell you it is the "study of numbers," "a lot of rules," or "a list of rules and procedures that need to be remembered." In contrast, if you ask mathematicians the same thing, they will tell you that mathematics is "the study of patterns" or "a set of connected ideas."

Moreover, there is confusion and debate about how to teach the "basics." How one teaches these to students has sparked much debate about pedagogical methodology. Mathematics teaching can be thought of in terms of two paradigms, a traditionalist approach, which suggests the practice of fundamental skills, and a reformist approach, which suggests discovery learning in the classroom to help students construct their own knowledge (Schoenfeld 2004; Karney et al. 2017). Alan Schoenfeld famously framed this debate in 2004 as the "math wars." In Schoenfeld's seminal article he articulated that the "math wars" are not simply about how one teaches math (traditionalist or reformist) but whether mathematics is for the elite or the masses, and thus involves tensions between "excellence" and "equity," and whether mathematics is a democratizing force or a vehicle for maintaining the status quo. Artfully, Schoenfeld (2004) emphasized why this conversation is of such importance—"mathematical knowledge, is a powerful vehicle for social access and social mobility. [The] lack of access to mathematics is a barrier—a barrier that leaves people socially and economically disenfranchised."

While these wars show no particular signs of nearing a resolution, some (Bevevino et al. 1999; Schoenfeld 2004; Carlson 2014) argue for a middle ground between teaching tried-and-true basic mathematics skills in the classroom and allowing students to explore, discover, and innovate. There is not a one-size-fits-all approach to teaching mathematics, or any subject. Below, I discuss some of the teaching methods, none of which intend to privilege one method over the other, but to demonstrate the strengths and challenges of each of them.

Rote Rehearsal and Deliberate Practice

Intrinsically we know what rote rehearsal (learning) is. It is based on repeating a task mnemonically, ingraining a skill, and thus developing the sense of gaining expertise from this repetition. The American Psychological Association (2020) provides a clear, concise definition, stating that the act of rote rehearsal or maintenance rehearsal involves "repeating items over and over to maintain them in short-term memory. According to the levels-of-processing model of memory, maintenance rehearsal does not effectively promote long-term retention because it involves little elaboration of the information to be remembered." Cramming for a test to have short-term retention of facts might make sense in some situations, but for deep learning rote rehearsal may not be an effective solution (Devlin 2019). Teaching "rote" suggests students learn blindly to follow rules and procedures with no foundational knowledge as to how the rules apply or why the procedure works (Mighton 2019).

A relative of rote learning is deliberate practice. Deliberate practice goes beyond the notion of repetition, creating meaning between a skill and its abstract goal—the goal is to build metacognitive proficiency. Made famous by K. Anders Ericsson and popularized by Malcolm Gladwell in *Outliers: The story of success,* deliberate practice suggests that students become experts through training, which is "a limited time of intense concentration and focused engagement, rather than simply rote rehearsal. The general characteristics of deliberate practice are setting specific goals, designing/monitoring learning activities, and reflecting" (Ericsson 2002; Gladwell 2008). Designing activities to promote discovery learning in mathematics, involves the use of problem sets that help students build connections and scaffolding to other facets of mathematics and other courses (i.e., fractals in biology). This means encouraging students to set goals for themselves that push the boundaries of their comfort zones, getting them to reflect upon the presented material. While there seems to be an ideological gap between deliberate practice and rote rehearsal, they actually are complementary. We need both. So, why do these two approaches seem to be polarized?

Discovery-Based Learning

In the debate, one side advocates the repetition of tasks and the other side supports process learning and problem solving, where students are responsible for constructing their own knowledge by developing their own explanations and approaches with little guidance from their instructor. As opposed to learning a single skill through the deliberate practice style of learning, discovery-based learning aims for a more holistic understanding of the process (Schoenfeld 2004). This might involve much struggle on the part of students.

The struggle is productive, however and is part of a method that "encourages creativity and builds authentic engagement and perseverance" (Cowen 2016).

However, since this is sometimes seen as a barrier to classroom learning, shifting teaching in a manner that promotes efficiency and correctness learning might be a remedy (Pasquale 2015). In this way, students can take ownership of their learning, gaining a better understanding that they might also transfer to other courses (Edwards 2018)—helping promote deep learning through the "discovery" of mathematics. However, discovery-based learning is also criticized because it might entail too much struggle, so that students might give up (Mighton 2019). Finding a balance is what is needed, just enough so that students are neither bored nor feel overwhelmed. This challenge-skill balance is termed *flow* by Csikszentmihalyi (2008), which "postulates that people enjoy an activity the most when the high challenge level of the activity matches people's high skill level." If the struggle is too great, students may be unmotivated, or worse intimidated, by the math problem in front of them.

Discovery-based learning has elements that inspire deep learning, promoting the flow; however, when scaffolding is absent and the struggle is too great students give up. This leads students to claim, "I'm just not a math person." What if instead there was a middle ground, one that introduces the practice (even through repetition), allowing students to "discover" how the math is connected, and instills confidence, and even joy?

Structured Inquiry

John Mighton (2020) offers a middle ground to close the gap between the traditionalists and the reformists. He describes his approach as "structured inquiry," a way to strike a "balance between independent and guided thought and between problems that are too hard or too easy for students." By focusing on incremental steps and providing the "right" scaffolding one can instill confidence in students as they move from simple to more complex mathematics. Others also encourage the use of this middle ground, blending tried-and-true basic mathematical pedagogy, allowing students to explore, discover, and innovate thoughtfully (Bevevino et al. 1999; Schoenfeld 2004; Carlson 2014).

This teaching approach, which combines task analysis and scaffolding, is a gateway to computer programming (coding) and algorithmic thinking. Task analysis, according to Mighton (2019), includes planning out the steps (down to minutiae) to follow and perform a procedure. This is an early foundation for algorithmic thinking, solving problems by following a set of rules, such as the rules of a calculation (Gollish 2019), typically associated with computer programing and the foundations of mathematical thinking. Deconstructing a mathematics problem according to its constituent steps can allow students to identify where they are struggling, so that an instructor can provide the right type help the student needs, as opposed to providing the solution or over-scaffolding the problem. This type of task analysis and writing out the steps follows George Pólya's classic method in *How to solve it* (Pólya 1957), in which he suggests four simple steps to break down and grasp any problem.

1. Understand the problem.
2. Devise a plan (translate).
3. Carry out the plan (solve).
4. Look back (reflect, interpret, and check).

In Ontario, this blended approach forms the basic pedagogy (Ministry of Education of Ontario 2005). Interestingly, Mighton's structured inquiry approach is sometimes criticized for merely giving a different name to "rote" or "drill and kill" learning. However, Mighton responds that by connecting a "basic" skill to a bigger idea this method neatly blends traditional and reformed approaches: "Most researchers now recommend that teachers introduce concepts with simple concrete models or representations and that they gradually make the representations more abstract" (Mighton 2020). Mighton shows how this method works with his lesson modules as part of the JUMP math program. While structured inquiry has its roots in an amalgam of deliberate practice and discovery-based learning, there are, however, other ways of teaching that can instill a passion for learning and for having fun with mathematics.

Other Ways to Teach Mathematics

Ultimately, the goal of math pedagogy is to teach students to thinking mathematically. Jordan Ellenberg (2014) suggests that mathematical thinking is an extension of common sense thinking, providing a lens to examine the world and useful for guiding informed decisions. Ellenberg advocates using the mathematical information that is present in the world, such as the occurrence of fractals in nature to stimulate mathematical thinking; in a similar vein Eugenia Cheng (2016) suggests that baking techniques and patterns can be used to simulate processes of abstraction.

Another approach is the so-called Habits of Mind one, which aims to stimulate different "skills, attitudes, cues, past experiences and proclivities" to reach a more thoughtful solution (Costa 2000). Developed by Costa and Kallick (2008), the goal in this method is to evoke attributes of "intelligent behavior." It posits that intelligence is not "fixed" but flexible, and with the right motivation anyone can learn anything (Costa and Kallick 2008; Boaler 2016).

Another method to teach mathematics is puzzle-based learning. Puzzles have one solution but multiple pathways to arrive at it; in effect it has attributes of both open-ended and closed-ended questions: "When students engage in puzzle-based learning, they explore facets of a task and begin to formulate a deeper understanding of the function which can be elaborated in class" (Costa 2017). This echoes learning and pedagogical principles such as those put forth by Pólya and Mighton. The ultimate goal is, again, to bring about deep learning in all math students, regardless of whether they show an aptitude for mathematics or not (Mighton 2008; Dweck 2008; Boaler 2016). Interestingly, none of these teaching methods suggest the use of devices such as calculators or computing devices; and they all assert that it is critical to not cognitively overload the students. Before discussing cognitive load, there are lessons to be learned outside of mathematics that might provide insights on teaching the basics of mathematics.

Literacy

The teaching of reading, literacy, music, etc. might offer ideas for math pedagogy. Literacy training is particularly relevant. As UNESCO (2002) put it, literacy today is a broad term: "Literacy is about more than reading or writing—it is about how we communicate in society. It is about social practices and relationships, about knowledge, language and culture" (Matsuura 2002). The Province of Alberta (Alberta Education 2020) provides a similar definition, remarking that literacy is: "the ability, confidence and willingness to engage with language to acquire, construct and communicate meaning in all aspects of daily living." Literacy is an acquired skill imparted through some form of training. Pre-literacy is, instead, the period in childhood when language is acquired without formal training. As is well known, as early as 6-months babies can recognize words, and by 8- or 9-months they can produce some words on their own, together with other communicative signs, such as gesture (Cicerchia 2020). Children soon develop the ability to identify letters, numbers, and shapes, as they develop phonemic competence and are taught how to associate phonemes with reading signs and systems; in English, this includes reading from left to right, and starting at the top of a page and reading down (Scholastic Parents Staff 2020). This pre-literacy stage forms the foundation on which children gradually learn to read and write.

Pre-literacy learning has always had implications for how to teach mathematics, since it might mirror how children develop numeracy alongside literacy. Some argue that the key process is making connections between the phonemic code and the alphabetic code, called phonics; others suggest instead that learning to read and write involves a focus on entire texts, rather than isolated words and their sounds, called holistic (Aarnoutse et al. 2001; Wexler 2018). The debate between phonics-versus-holistic learning of literacy debate has migrated to the elementary classroom, where mathematics and reading are taught in tandem. Educators are coming to recognize that a balanced approach is the most effective one to teach reading, and this extends to mathematics, whereby symbols and concepts can be connected either as discrete items (as in phonics) or as part of a theme in which mathematical concepts can be located (as in the holistic approach) (Davis and Mighton 2018).

In all this, motivation is a key factor. Research indicates that literacy and numeracy improve when a student develops a love for reading and math. This means contextualizing learning; so, for example, getting a student who is interested in baseball to read a passage about baseball the passage might enhance the student's comprehension and analytical skills; the same applies to doing, say, a puzzle in math or working with practical math problems (Mighton 2008; Wexler 2019). In one of her articles Wexler (2019) asks:

What if the best way to boost reading comprehension is not to drill kids on discrete skills but to teach them, as early as possible, the very things we've marginalized—including history, science, and other content that could build the knowledge and vocabulary they need to understand both written texts and the world around them?

From phonics to building a mental lexicon of familiar words, how one gets from one word at a time (the pre-literacy phase) to full reading comprehension across a variety of subjects does not matter—it is getting the student there (Cicerchia 2020).

Physical Literacy

Physical literacy refers to the acquisition of skill in one or more sports. It involves knowing how to move physically in a specific sport. Overall, it involves knowledge of the fundamental movements and skills in any kind of physical activity (Kriellaars 2013). Learning to be physically literate it is not, however, merely knowing how to move in an environment of play, but also how to carry out new activities, how to take advantage of opportunities for working together, and how to ask relevant questions (Kriellaars 2013; Sport for Life 2020). Children need to be taught agility, balance, and coordination in different environments, inside and outside. Confidence and competence come from enjoyment, repetition, and the safe testing of the child's limits to improve their abilities. Sport for Canada advocated a minimum of 180 minutes (3 hours) of activity per day for children, with 60 of those being vigorous.

Physical literacy programs provide activities for children during their school years and even beyond. The fundamental phases that undergird such programs are the following (Higgs et al. 2019):

1. *Active Start*: this is designed to help children master basic movements and develop habits of physical activity through fun, engaging activities.
2. *FUNdamentals*: this is intended to help children develop fundamental movement skills—agility, balance and coordination—again through the enjoyment of physical activity.
3. *Learn to Train:* is the phase when children learn a wide variety of sport-specific skills in an enjoyable and friendly environment.

The emphasis on "fun" is a key principle in this approach, as is the concept of general understanding. In his book *Range: Why generalists triumph in a specialized world*, David Epstein (2019) exposes the myth behind early specialization: "eventual elites typically play a variety of sports, usually in an unstructured or lightly structured environment; they gain a range of physical proficiencies from which they can draw." Epstein compares the career paths of Tiger Woods and Roger Federer in their respective sports (golf and tennis). Although Woods appears to have been exposed to a highly structured routine from a young age, both athletes enjoyed similar success because they enjoyed practicing their sports, which allowed them to develop proficiency in them.

In 2015, Sport for Life defined physical literacy as "the motivation, confidence, physical competence, knowledge and understanding to value and take responsibility for engagement in physical activities for life," based on affective (motivation and confidence), physical (physical competence), cognitive (knowledge and under-standing), and behavioural (engagement in physical activities for life) modalities. It identified five core principles of physical literacy that, on closer scrutiny, apply to any form of literacy:

1. It is an inclusive skill accessible to all.
2. It represents a unique journey for each individual.
3. It can be cultivated and enjoyed through a range of experiences in different environments and contexts.
4. It needs to be valued and nurtured throughout life.
5. It contributes to the development of the whole person.

The Musical Mind

Music might be the most similar to mathematics in its structure and in the way the two are acquired (Ericsson 2002; Gladwell 2008). Musical literacy is thus also relevant to math pedagogy. Music concepts are abstract, so they are best taught with visuals and manipulatives at first (Alegria 2017). Music also requires dedicated practice to master a new skill, as does mathematics. It too can have "hugely positive ramifications for personal fulfillment and lifetime success" (Tsioulcas 2012; Buszard 2014). For these reasons, music and mathematics pedagogy share many features. Some of the classic methods, such as the Suzuki and Kodály one, have in fact been extended to mathematics (Sarrazin 2012). It is worth going through the main ones for the sake of illustration:

1. *The Kodály Method* is a holistic approach focusing on the intellectual, emotional, physical, social, and aesthetic aspects of music, with the belief that music is for everyone. It espouses a sequential approach, beginning with sight-reading and mastering basic rhythms and pitches before advancing to complex technical and aesthetic aspects.
2. *The Dalcroze Method*, also called the eurythmics method, focuses on teaching rhythm, structure, and musical expression through movement. In this approach the music is the stimulus for making the body move, which in turn causes an emotional reaction in the student, deepening the significance of the experience. The method begins with training the ear so as to get the body to mimic the sounds.
3. *The Orff Schulwerk Approach* fosters creative thinking through improvisation, combining instruments, singing, movements and speech to develop musical proficiency. There are four phases in this approach: imitation, exploration, improvisation and composition.
4. *The Suzuki Method* espouses principles that are meant to match how children acquire their native languages; a learner begins by listening and then repeating in

a step-by-step process, inducing patterns in the music and then practicing them formally.

As mentioned, these methods have much in common with the ways in which mathematics is learned and taught. So, rote rehearsal could be compared to decoding words in pre-literacy environments (reading), consistent repetition in sport with practice techniques in mathematics, and discovery-learning (induction) in the Suzuki Method (music) corresponds to structured learning; and so on.

Cognitive Load

In all kinds of learning tasks, carrying a huge cognitive load will invariably hamper learning. Cognitive load refers to the working memory (short-term) required to carry out a task (Kirschner et al. 2018). The more a task demands of us cognitively, the slower we learn. Most of us now probably do not need much working memory or time to do the times tables between 1 and 12 (these are in our long-term memory); but learning them initially involved a considerable cognitive load (mnemonic effort). In addition to memory, cognitive load involves the number of tasks one is trying to complete. "At high levels of multitasking, the cognitive load is higher and the benefits smaller. It is modern technology and computers that allow people to multitask—for instance, web-browsers provide an interface that allows for multiple tabs to facilitate concurrent activities" (Adler and Benbunan-Fich 2012).

This is where the notion of *switching cost* comes in. When we multitask, instead of solo- or single-tasking, there is a cognitive cost to pay, whereby we must shift our brain states between the concurrent tasks we are attempting to complete. It is almost impossible to reach a state of *flow* if we are multi-tasking (American Psychological Association 2006; Aral, Brynjolfsson, and can Alstyne 2011; Adler and Benbunan-Fich 2012).

Conclusions

To argue for a no-technology, machine-free, classroom today is an absurdity, given that we are immersed in all kinds of technologies. If nothing else, technology should be used as an ancillary tool for reinforcing learning. However, if the students become solely focused on the use of devices for all tasks there is a switching cost involved, disrupting the thinking flow. In a seminar organized by *Science Magazine* (Oransky et al. 2019), it was pointed out, virtually by consensus, that it is important for students to understand basic mathematics independently, and not "relying and overlying on software," because "it can lead to trouble." The trouble is lack of understanding and misuse of basic mathematics. Technology can be seen as a scaffold with respect to traditional teaching.

To make sure that the basics are learned, a balanced approach is required, based as well on insights from other subjects. We cannot separate graduated material, rote practice from discovery-induction. We must also learn from methods used to impart other literacies, as discussed. If I were to dig down to the granularity of what I think the fundamentals of mathematics are, I would see them as components of a clock. Similar to a clock there are 12 numbers around the face, but instead of hands in the middle, there are the four operations—addition (+), subtraction (−), multiplication (×) and division (÷). All students should be able to not just complete these operations from 1 to 12, but also understand what the operations mean and be able use them as part of their daily experiences. These basic operations parallel the *FUNdamentals* of physical literacy; playing with numbers from 1–12 is the *FUNdamental* of mathematics. My claim is that through these simple pedagogical ideas, we can bring back the "fun" to the math classroom.

References

Aarnoutse, C., Van Leeuwe, J., Voeten, M., and Han, O. (2001). Development of decoding, reading comprehension, vocabulary and spelling during the elementary school years. *Reading and Writing: An Interdisciplinary Journal* 24: 61–89.

Adler, R. F. and Benbunan-Fich, R. (2012). Juggling on a high wire: Multitasking effects on performance. *International Journal of Human-Computer Studies* 70: 156–168.

Ahlfors, L. V., Bacon, H. M., Bell, C. V., Bellman, R. E., Bers, L., Birkhoff, G., and Garabedian, P. R. (1962). On the mathematics curriculum of the high school source. *The American Mathematical Monthly* 69: 189–193.

Alberta Education. (2020). *Literacy and numeracy.* https://education.alberta.ca.

Alegria, M. (2017). *Music as a teaching tool: Incorporating music into almost any class can be a great way to teach content—and it doesn't take special training or expensive resources.* https://www.edutopia.org/

American Psychological Association (2006). Multitasking: Switching costs. *American Psychological Association.*

American Psychological Association (2020). *Maintenance rehearsal.* https://dictionary.apa.org/maintenance-rehearsal.

Aral, S., Brynjolfsson, E., and Van Alstyne, M. (2011). Information, technology and information worker productivity. *Information Systems Research*, SSRN: https://ssrn.com/abstract=942310.

Bevevino, M. M., Dengel, J., and Adams, K. (1999). Constructivist theory in the classroom: Internalizing concepts through inquiry learning. *The Clearning House* 72: 275–278.

Boaler, J. (2015). *What's math Got to do with It? How teachers and parents can transform mathematics learning and inspire success.* New York: Penguin Books.

Boaler, J. (2016). *Mathematical mindsets.* San Francisco: John Wiley and Sons.

Buszard, T. (2014). The secrets of self-taught, high-performing musicians. *The Conversation.* https://theconversation.com.

Carlson, K. B. (2014). *Math wars: The division over how to improve test scores.* http://www.theglobeandmail.com/news/national/education/making-the-grade-the-provincial-push-for-better-math-scores/article16290351/.

Cheng, E. (2016). *How to bake pi: An edible exploration of mathematics.* New York: Basic Books.

Cicerchia, M. (2020). *Read and spell.* https://www.readandspell.com.

Costa, A. L. (2000). *Habits of mind.* Sacremento: The Institute for Habits of Mind.

Costa, A. L. and Kallick, B. (2008). *Learning and leading with Habits of Mind: 16 essential characteristics for success.* Alexandria: Association for Supervision and Curriculum Development.

Costa, S. (2017). Puzzle-based learning: An Approach to creativity, design thinking and problem solving: Implications for engineering edcuation. *Proceedings of the 2017 Canadian Engineering Education Association (CEEA17) Conf.* Toronto: CEEA.

Council of Ministers of Education, Canada. (2020). *Literacy at CMEC.* https://www.cmec.ca.

Cowen, E. (2016). Harnessing the Power of Productive Struggle. *Edutopia: The George Lucas Educational Foundation* .

Csikszentmihalyi, M. (2008). *Flow: The psychology of optimal experience.* New York: Harper Perennial Modern Classics.

Davis, B. and Mighton, J. (2018). In the continuing 'math wars,' both sides have a point. *The Globe and Mail*, 2018.

Devlin, K. (2019). How technology has changed what it means to think mathematically. In: M. Danesi (ed.), *Interdisciplinary perspectives on math cognition.* Cham: Springer Nature Switzerland.

Dweck, C. S. (2008). *Mindset: The new psychology of success.* New York: Ballantine Books.

Edwards, C. (2018). Productive struggle. *National Council of Teachers of Mathematics*.

Ellenberg, J. (2014). *How not to be wrong: The power of mathematical thinking.* New York: The Penguin Press.

Epstein, D. (2019). *Range: Why generalists triumph in a specialized world.* New York: Riverhead Books.

Ericsson, K. A. (2002). Attaining excellence through deliberate practice: Insights from the study of expert performance. In: M. Ferrari (ed.), *The pursuit of excellence through education*, 21–43. Mahwah: Lawrence Erlbaum Associates.

Finkel, D. (2019). How can play help us understand math? (G. Raz, Interviewer) TED: Radio Hour.

Gladwell, M. (2008). *Outliers: The story of success.* Boston: Little, Brown and Company.

Gollish, S. (2019). *An investigation into mathematics for undergraduate engineering education to improve student competence in important mathematics skills.* Toronto: University of Toronto.

Higgs, C., Way, R., Harber, V., Jurbala, P., and Balyi, I. (2019). *Long-term development in sport and physical activity 3.0.* Sport For Life. Canada.

Karney, B., Mather, A., and Gollish, S. (2017). A decision-making framework for engineering mathematics education: A critical literature review. *Conference Proceedings: Canadian Engineering Education Association (CEEA2017)* (Paper 61). Toronto: CEEA.

Khan Academy. (2020, 05 19). *Early math.* https://www.khanacademy.org/math/early-math.

Kirschner, P. A., Sweller, J., Kirschner, F., and Zambrano, J. R. (2018). From cognitive load theory to collaborative cognitive load theory. *International Journal of Computational Support and Collaborative Learning* 13: 213–233.

Kriellaars, D. (2013). *Physical literacy assessment for youth.* Victoria: Canadian Sport Institute—Pacific.

Merriam Webster (2017). *Mathematics.* http://www.merriam-webster.com/dictionary/motivation.

Merriam-Webster (2020). *Definition of fundamental.* https://www.merriam-webster.com/dictionary/fundamentals.

Mighton, J. (2008). *The end of ignorance: Multiplying our human potential.* Toronto: Vintage Canada.

Mighton, J. (2019). Using evidence to close the achievement gap in math. In: M. Danesi (ed.), *Interdisciplinary perspectives on math cognition*, 265–276. Cham: Springer Nature Switzerland.

Mighton, J. (2020). *All things being equal: Why math is the key to a better world.* Toronto: Knopf Canada:

Ministry of Education of Ontario (2005). *The Ontario curriculum grades 1-8 Ministry of Education mathematics.* Toronto: Government of Ontario.

National Centre for Education Statistics (2020). *Numeracy domain.* Washington: Department of Education.

Oransky, I., Harris, R., Scott, C. T., and Jasny, B. (2019). Fighting fake science: Barriers and solutions. American Association for the Advancement of Science (AAAS).

Organization for Economic Cooperation and Development (2019). *PISA 2021 Mathematics framework*. Paris: OECD.

Pasquale, M. (2015). *Productive struggle in mathematics*. Waltham: Education Development Center.

Pólya, G. (1957). *How to solve it: A new aspect of mathematical method*. Princeton: Princeton University Press.

Sarrazin, N. (2012). *Music and the child*. Buffalo: SUNY Textbooks.

Schoenfeld, A. H. (2004). The math wars. *Educational Policy* 18: 253–286.

Scholastic Parents Staff. (2020). *The meaning of preliteracy: How your child's early experiences shape his learning success*. https://www.scholastic.com.

Sport for Life. (2020). *Key factors underlying long-term development in sport and physical activity*. https://sportforlife.ca/.

Tsioulcas, A. (2012). *Getting kids to practice music without tears or tantrums*. https://www.npr.org/sections/deceptivecadence/.

UNESCO. (2002). *United Nations literacy decade: Education for all*. Paris: UNESCO.

Vinner, S. (2018). *Mathematics, education, and other endangered species: From intuition to inhibition*. Cham: Springer Nature Switzerland.

Wexler, N. (2018). *Why American students haven't gotten better at reading in 20 years*. https://www.theatlantic.com/.

Wexler, N. (2019). Natalie Wexler on America's knowledge gap. *Amplify*. go.info.amplify.com.

Chapter 10
Syntonic Appropriation for Growth in Mathematical Understanding: An Argument for Curated Robotics Experiences

Krista Francis and Steven Khan

Introduction: "There Are No Numbers Between 5 and 6"

We open with an anecdote from our initial co-writing meetings about students' growth in mathematical understanding. We chose this scene to provide an illustrative example of what we intend to highlight, viz. that curated robotics experiences offer opportunities for syntonic appropriation that contributes to learners' growth in mathematical understanding:

Krista: When I work with children aged 9–11 years old (Grade 4 to 6) with robots I've noticed in their conversations they often refer to the robot as themselves ("I need...") or extensions of themselves ("We need…"). Adults, teachers and pre-service teachers, also often refer to the robot as themselves. For example, when they say "we're *so* close" as they try to get the robot to travel 100 centimeters exactly or "we need to go back one" on similar tasks. I've also noticed how quickly children seem to learn decimal numbers when they do this task or similar tasks. When I first started working in one school (3 years ago) and asked what's between 2 and 3 or 4 and 5, it didn't matter which grades, 4, 5 and 6 (Division 2 or Upper Elementary) the children would confidently answer "there is nothing" between 2 and 3 or between 4 and 5 and that's despite having up to three years of learning about decimal numbers.

Steven: Since I work mostly with elementary pre-service teachers, let's look at the K-8 Program of Studies (Alberta Education 2016a, b) (Curriculum document) and the Achievement Indicators (2016) (Supplementary document) to see where decimals first appear, as what happens in classrooms here in Canada I've found is very much driven by the Provincial Curriculum documents. Decimals appear in Grade 4 in the Number Strand (see Appendix) and the goal of comparing and ordering decimal numbers becomes a required curriculum outcome by the end of Grade 5. The specific model of using a number line is not

K. Francis (✉)
Werklund School of Education, University of Calgary, Calgary, Canada
e-mail: kfrancis@ucalgary.ca

S. Khan
Brock University, St. Catharines, Canada
e-mail: skhan6@brocku.ca

© Springer Nature Switzerland AG 2020
S. A. Costa et al. (eds.), *Mathematics (Education) in the Information Age*,
Mathematics in Mind, https://doi.org/10.1007/978-3-030-59177-9_10

mentioned in the Program of Studies but is in the accompanying Achievement Indicators document which is meant to provide some ideas to teachers but is not the 'curriculum' and teachers are not legally required to report on the ideas presented there.

Krista: Grade 4 teachers might say that this (using the number line for decimals) is not their responsibility. But the number line is such a powerful tool (Braithwaite and Siegler 2018; Obersteiner et al. 2019). It elaborates and extends the ordering of whole numbers, especially for decimal fractions, and points towards the real number continuum later on that it is surprising to me that it is not even considered until Grade 5.

Steven: Checking the Program of Studies, I see that the number line is introduced as an expectation in Grade 1 with respect to benchmarks of the whole numbers 0, 5, 10 and 20 but is not with respect to fractions and decimals at least not in the stated expectations. It is however explicitly called on in the Achievement Indicators in Grade 4 (Name fractions between two given benchmarks on a numberline. Order a given set of fractions by placing them on a number line with given benchmarks) and Grade 5 (Position a given set of fractions with like and unlike denominators on a number line, and explain strategies used to determine the order; Order a given set of decimals by placing them on a number line that contains the benchmarks 0.0, 0.5 and 1.0.) and Grade 6 (Place a given set of fractions, including mixed numbers and improper fractions, on a number line, and explain strategies used to determine position.)

Krista: Well, I am going to keep this information in mind for future teaching and professional learning. Below (Fig. 10.1) is an illustration of an interaction that two Grade 4 girls had with their teacher while programming their robot to travel exactly 100 cm. This was their first experience programming the robot to move and their first experience with decimal numbers. They had been working on this task for about half an hour before this interaction occurred. They had observed that 5-wheel rotations did not travel far enough and 6 was too far. Their challenge was understanding that there were numbers between 5 and 6.

Krista: The next week this pair of girls were using decimal numbers as they were trying to figure out how many wheel rotations to travel 73 cm. They solved the questions with an answer of 4.2. This was a variation of the context and the girls demonstrated a spatial sense of the number indicating it was closer to 4 than 5.

We framed our discussions about these incidents in terms of a growth in mathematical understanding following the model offered by Pirie and Kieren (1994a, b). We believe that a well structured robotics inquiry (such as these described in this paper and others) allows students to discern critical features of a concept (Marton 2014, 2018) through providing multiple instantiations of the concept (available through different embodied metaphors and enactions) and multiple opportunities to relate to its different aspects.

In this paper we argue that a well structured robotics inquiry can lead to what Pirie and Kieren (1994a, b) called growth in mathematical understanding. In particular we offer that such structuring is a means to encourage processes of syntonic appropriation as introduced by Papert (1980). We start with the observation that some mathematical concepts are introduced to learners in ways that are disassociated from learners' bodies, experiences and/or culture(s). Consequently, learners struggle to apply/relate the mathematics concepts in novel circumstances – such as in a robotics environment. Learning in this case is superficial and fragmented though our intention is for such knowledge to become deep and connected.

Fig. 10.1 How many wheel rotations for the robot to travel 100 cm?

The intentional design of instructional tasks can focus use of mathematical concepts for making the robots move more precisely in order to prompt growth in mathematical understanding around mathematical concepts such as the existence of rational numbers and their 'location' or 'magnitude' on the visuo-spatial representation of a number line. Our goal is to suggest and illustrate how the Pirie-Kieren model can be used to highlight/draw attention to some of the growth in mathematical understanding within a curated robotics learning experience. The growth in mathematical understanding we observe involves students (gradually) appropriating models and concepts in a syntonic way through curating their own experiences of learning from the opportunities provided by the teacher's previously curated robotics learning task(s). We use the metaphor of curating in this paper as we have found that there is value in shifting our thinking from the language of *accumulating* of experience (Khan et al. 2015) to the intentional and deliberate *curating* of such experience which involves keeping an imagined audience and their interactions in mind.

Arguing for Curating Tasks—Theoretical Framing

Knowing Is Doing

Our main theoretical commitment is to enactivism (see Khan et al. 2015). However, in this paper we also draw on theories of embodied cognition in the learning of mathematics (Lakoff and Núñez 2000) and computational thinking (Buteau et al. 2016; Francis et al. 2016; Grover and Pea 2013; Wing 2006).

Briefly, enactivism is (1) a theory of engagement, (2) that is simultaneously attentive to the coupling of organisms and their environments, action as cognition, and sensorimotor coordination; and (3) attending to relevant phenomena of interest involves a methodological eclecticism (Di Jaegher and Di Paulo 2013) that is concerned with inter-agent dynamics that include feedback from the system and the organism's responses. In our work students, teachers, researchers, tasks and technologies are dynamically coupled and provide feedback to each other.

Enactivist theories of human learning attend explicitly and deliberately to action, feedback, and discernment. They emphasize the bodily basis of meaning. As Brown and Coles (2011: 861) note, "[t]he enactive conception of knowledge is essentially performative", i.e. knowing is doing. While constructivism can also be interpreted as performative, the focus is on the outcome of actions rather than the process of interactions as in enactivism. Enactivism is attentive to the many feedback structures in a greater-than-the-individual-learner system. It is the organism as a whole, together with its environment, which co-evolves in enactivism.

We work from the position of Varela et al. (1991: 173) that contends that the enactivist approach comprises two principles, viz. that "(1) perception consists in perceptually guided action and (2) cognitive structures emerge from the recurrent sensorimotor patterns that enable action to be perceptually guided." (That is to say, what an individual perceives is dependent on, but not determined by, the types of sensory stimuli that the individual's body, its physical interface with the world encounters.

For example, with respect to perceptually guided action, Rushton (2008: 36) states, "[t]o walk to a target you need to know where it is," i.e., our potential for action (walking or moving in terms of the robot and body syntonicity) and goal (destination) is dependent on the perception and selection of sensory information from the physical world. Work in developmental psychology (e.g. Keen et al. 2003), which analyses infants reaching for objects, exemplify both principles of the enactivist approach we are using.

Over time, repeated activity (action) establishes predictable (statistical) patterns of neural and neuromuscular activity that in turn influences the response to the stimuli sought or encountered. Humphreys et al. (2010: 186) argue and present experimental evidence that, "our need to act upon the world not only imposes a general need for selection on our perceptual systems, but it directly mediates how selection operates. Attention is grounded in action." It is action then or rather potential for action that focuses attention on some features of the environment such that some aspects of the sensory landscape are perceived and others are not. It is these ongoing focusing of perceptions and sensations and evaluation of goal states (feedback) that guides action. This focusing of intention is an attribute of the individual learner as well as the learning designer-teacher-curator.

We distinguish between related enactivist and embodied perspectives in that in enactivism the external environment plays a significant role in understanding the dynamic unfolding of cognitive processes: what is in the environment is a resource for thinking, doing/knowing and being. While enactivism is attentive to ongoing co-constituted interaction among bodily action, cognition, and the environment, theories of embodied cognition focus on the relationship between cognition and *prior* action. In other words, cognition, in embodied models, is closely tied to prior sensorimotor experiences. Rather than predictions of learning a concept, enactivism is concerned with the learning *in* action since it is the potential for action in the world that focuses attention and drives learning. Embodied cognition is concerned with the learning *from* action. Embodied cognition can be regarded as a sort of consolidation of enactive action. In this work both of these theoretical frames are necessary in our attempt to make sense of learners' growth in mathematical understanding in a robotics learning environment.

Hutto (2013: 174) argued that enactivism, with its starting assumption that mental life can be understood as embodied activity, is a good candidate for "defining and demarcating [psychology's] subject matter"—that is, in his terms, for "unifying psychology." Traditional perspectives, he argued, delimit psychological explanations to ones that rely on inner representational states. He noted that enactivism, in its original formulation by Varela et al. (1991: 177), attended explicitly to organisms' varied engagements with contexts "not only of the biological kind but also of sociocultural varieties." The robotic moving task illustrated in Fig. 10.1 could be interpreted as merely a manipulation of an inner representation but to our understanding from an enactivist perspective it is not at this point in time. Rather, they are in the process of constructing a mental number line (as an *object-to-think-with*) through action with the robot, the programming interface, the task specifications, and each other. By the last panel though where they have completed the task accurately, embodied perspectives help us to better understand how decimal numbers and a number line have been appropriated through metaphors of imagined robotic movement. We see value here in drawing readers' attention to this necessary shift in theoretical tools to better analyse and understand the growth in mathematical understanding of learners at two different points in developmental time. Bridging these two moments in time we use the Pirie-Kieren model which helps us to carefully notice and name subtle shifts in attention as evidenced by changes in language, gesture and performance (action).

Teaching Is Presenting Appropriable Challenges

From Accumulating to Curating

In previous work we used the idea of learning from an enactivist perspective as an *accumulating* of sufficient and diverse experiences (Francis et al. 2016; Khan et al. 2015). Over time, in work with students and teachers the limits of this descriptor have become more apparent as we have critically appraised our own growth in understanding about how people (National Academies of Sciences, Engineering, and Medicine 2018) and systems learn (Davis et al. 2019; Dehaene 2020). Learners and teachers do appreciatively much more than 'accumulate' experiences, they attach affective (probabilistic) weights and meaning to these experiences. The metaphor of curating has emerged as a more apt descriptor than accumulating.

We take curating as a literacy practice. Looking to its linguistic origin we find both curation as a noun and a verb. As a noun, a curate refers to a person tasked with the care or cure of souls. We choose to read 'care' as deliberate and loving attention to the necessary aspects for the realization of well-being in another (including the self). We read 'cure' (of a soul) not in a medically restorative sense or the elimination of a disease, but rather as a learning how to live and be well in the world (Seligman 2011) with others. In more recent usage, the verb curate refers to the actions of selecting, organizing, and presenting something for an intended (or imagined) audience based on expert knowledge and values. According to art historian, Donald Preziosi (2019: 11), curation,

> involves the critical use of parts of the material environment both for constructing and deconstructing the premises, promises, and potential consequences of what are conventionally understood as realities, or social, cultural, political, philosophical, or religious truths. It is a way of using things to think with and to reckon with—to struggle with and against—their possible consequences. It is an epistemological technology: a craft of thinking. As such, it is not innocent or innocuous…[It] entails the conscious juxtaposition and orchestration of what in various Western traditions were distinguished as "subjects" and "objects": what are conventionally differentiated as "agents," and as what is "acted upon." Curating not only precedes and is more fundamental than exhibitions, galleries, collections, and museums; but it is also not unique, nor exclusive, to any of those institutions and professions. In fact, it is not even an "it" at all but is, rather, a way of using things: potentially any things. In short, curating is a creative performance using the world to think about, and both affirm and transform, the world.

Curation here involves critical, craft and creative thinking with awareness or consciousness, i.e. it is not mere collection (or accumulation) and display, it is oriented towards an imagined audience and intended for learning. The perspective above shares many resonances with Papert's framing of bricolage (Papert 1980). It is engaging in bricolage that provides the necessary and diverse occasions for syntonic appropriation.

As independent curator Glen Adamson argues, curation is about manipulating and trading in the attention economy, "You are drawing people's attention to objects in a different and heightened way. The other big idea for me is that curation is really about attention. The medium you work in as a curator is attention, and we live in a so-called 'attention economy,' in the sense that what people pay attention to is itself a form of value" (Acosta and Adamson 2017). Attention is one pillar that is key to learning (Dehaene 2020). Our goal is to suggest and illustrate how the Pirie-Kieren model can be used to highlight/draw attention to some of the growth in mathematical understanding within a curated robotics learning experience:

> *Krista*: When I first started doing this work my attention was on the engineering process (design), partly because I was taught/mentored in the design of robotics tasks but my son who was in engineering at the time. But as I started to recognize glimmers of the potential for mathematics learning, I began to curate and explore more mathematical tasks as opposed to design tasks. I found that as students gained some of the mathematical connections in programming robots, their skill at the programming design also improved, such as manoeuvring the robot precisely. Michael recently told me that if he would have learned how the move steering worked in Grade 9 (mathematically modeled) it would have put him years ahead in the robotics competitions in which he participated.

Rephrasing this in the language of our paper, a deeper more complex and focused syntonic appropriation and appreciation of the robot's functioning might occur alongside (in synchrony with) a growth in mathematical understanding needed for personal goal achievement. Getting there however requires our intentional focusing of students' attention through our curated task. Teachers, we think, have always worked the economics of attention in classrooms, schools and larger collectives, and in doing so have developed or utilised skills in curating. What we are trying to do is to draw attention to that challenging aspect (curating) of teachers' work that is not quite captured in the idea of Mathematics-for-Teaching (M4T) (Davis and Renert 2014) or Mathematical Knowledge for Teaching (Ball et al. 2005) or Technological Pedagogical Content Knowledge (TPCK) (Koehler and Mishra 2005) and to find ways to value it and develop it more intentionally as part of the work we do in our different professional networks with pre-service and in-service teachers and colleagues in different communities of practice:

> *Steven*: Teachers and pre-service teachers have taught me the value of the emotional dimension for learners in tasks, I hope that I manage to shift their understanding that while learners might be engaged because of the emotional investment in the task, their attention as a teacher has to be on what is mathematically significant.

In our next section, we introduce some elements of the Pirie-Kieren model of growth in Mathematical Understanding which we use as an analytic frame.

The Pirie-Kieren Model of Growth in Mathematical Understanding

There are a variety of framings of mathematical understanding (Hiebert and Carpenter 1992; Sfard 1991; Sierpinska 1994; Simon 2006; Skemp 1976). George (2017) offers a historical analysis of the concept in mathematics education. An enactivist framing of understanding however grounds it in terms of the dynamics of action (or co-action) and potential actions in a world, that is to say, individual understandings are not static or 'fixed' but contextually and temporally dependent, grounded in experience and interpretations of experience, that may be challenging to articulate. This view of understanding is a non-linear or complex one (Davis and Simmt 2003). As such, we draw on Pirie and Kieren's (1994a, b) model of growth of mathematical understanding through non-linear back-and-forth movements of the following modes: *primitive knowing, image making, image having, property noticing, formalising, observing, structuring* and *inventising* (see Fig. 10.2).

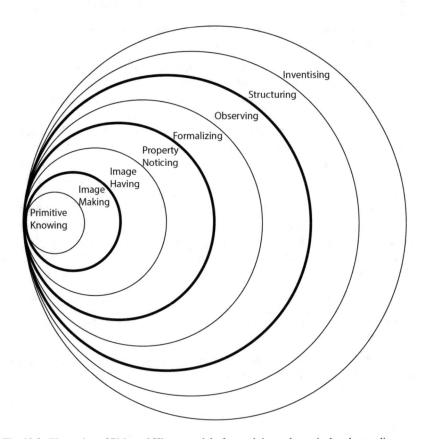

Fig. 10.2 Illustration of Pirie and Kieran model of growth in mathematical understanding

We use the Pirie-Kieren model as an analytic tool to illustrate children's growth in understanding of mathematical concepts and computational thinking concepts (see Namukasa 2019 for an example of the relationship). Coming to understanding starts with primitive knowing. "Primitive here does not imply low level mathematics, but is rather the starting place for growth of any particular mathematical understanding" (Pirie and Kieren 1994a: 170). This is what a learner brings with them at the beginning of a new task sequence. According to Pirie and Kieren (1994a), students' *primitive knowing* can be assumed, i.e., the skills they have initially. In the case of these students working with Lego EV3 robots such *primitive knowing* includes knowing related to mathematics such as spatial reasoning, proportional thinking, number; programming; technical procedural knowing such as how to connect the robots to the iPad and download and run programs; and knowledge about classroom routines, norms and procedures including how to work in small groups:

- *Image making* is when a student records and reflects on *primitive knowing* (creating an object through drawing or manipulatives). In figure (1) the girls are beginning to think about (make an image) of what is between the numbers 5 and 6 (prompted by their engagement with the robot as well as their teacher and the task).
- *Image having* is when a student no longer requires acting on the object. Pirie and Kieren (1994a: 170) note that, "[a]t the mode of *image having* a person can use a mental construct about a topic without having to do the particular activities which brought it about." Between the second and the third frame of Fig. 10.1 above, the students are starting to recognise how the decimal numbers are related incrementally, i.e. they are beginning to develop a spatial sense of (decimal) numbers on a number line.

At the fourth level of understanding, *property noticing*, one can manipulate or combine aspects of one's own images to "construct context specific properties" (Pirie and Kieren 1994a: 170). In our example, the fourth frame when students say "we need 7", we take this as indicating a movement from *image-making* to *image having* in that students are no longer thinking directly 'with' the robot but are able to use their mental image of decimals on a number line. Mathematically, that is as far as we perceived the students' movement in the model. Having also worked with Marton's (2014, 2018) Variation Theory of Learning we see the various modes in the PK model as dynamic networks of critical discernments of student thinking or understanding. This is particularly evident with the *property noticing* mode.

However, from a programming standpoint, in frame 2 of Fig. 10.1, the students very quickly moved from *primitive knowing* of how to program to move whole wheel rotations, through *image making, image having, property noticing* to *formalising* how to program decimals into the move steering block. *Formalising*, is when one can abstract a "method or common quality from the previous image" (Pirie and Kieren 1994a: 171). We have demonstrated previously (Francis et al. 2016) that the block programming environment makes it easier to move into this formalising mode.

Deeper levels of the Pirie and Kieren model (1994a: 171) are not observed in our analysis, but for reference they are mentioned briefly. *Observing* occurs when one reflects on one's *formalising* of images and proposes theorems. "*Structuring* occurs when one attempts to think about one's formal observations as a theory. This means that the person is aware of how a collection of theorems is interrelated and calls for justification or verification of statements through logical or meta-mathematical argument." With *inventising*, one has such a strong understanding that they are able to ask new questions which might grow into an entirely new concept. In our example in Fig. 10.1, the grade 4 girls do not yet have sufficient number or sufficient diversity in the example space of curated experiences in mathematics or robotics or programming to move into these levels/modes of understanding. That is to say, they are in the process of developing and growing their understandings in each of these areas.

Syntonic Appropriation and Tools for Conviviality

As we mentioned earlier, the teacher and students in Fig. 10.1 above referred to the robot in the first-person plural as 'we'. This use of the word 'we' is an indication of the intimacy, familiarity and conviviality they have with the robot. As a similar example, Merleau-Ponty (1978) noted the phenomenon of such intimacy, familiarity and conviviality with technology in the example of driving a car. Drivers are intimately acquainted with how the car turns by moving the steering wheel, or how pressing the gas pedal changes the speed, or how pressing the brake arrests movement. Drivers are also intimately aware of the dimensions of the car so that when they are parking, they do not bump into other cars. There is a sense of knowing and intention such that the driver moves the car with almost the same spatial precision as their own body (Merleau-Ponty 1978: 144):

> We said earlier that it is the body which "understands" in the acquisition of habituality. This way of putting it will appear absurd, if understanding is subsuming a sense datum under an idea, and if the body is an object. But the phenomenon of habituality is just what prompts us to revise our notion of "understand" and our notion of the body. To understand is to experience harmony between what we aim at and what is given, between the intention and the performance—and the body is our anchorage in the world.

The experience of harmony is part of what Papert (1980) intends by the term 'syntonic appropriation' and Illich (1973) by 'convivial.' Tool use is not an end in itself nor is it the motivation for action. Doing something, creating something and the aesthetic dimensions of experience are the ends and motives:

> the ultimate theoretical task in advancing, for example, the learning of mathematics, is not producing a range of so-and-so-centric kinds of mathematical knowledge but rather finding ways of thinking about mathematical knowledge that will allow each individual to make what in Mindstorms I call a syntonic appropriation (Papert 1980).

In *The children's machine,* Papert (1993) connects bricolage as a methodology for intellectual activity in the context of tinkering, building with Lego, and working in computer environments (programming in Logo and controlling robot turtles) with Illich's (1973) concept of tools for conviviality, seeing the latter as analogous to his concept of syntonic appropriation. Papert (1993: 144) writes:

> [t]he basic tenets of bricolage as a methodology for intellectual activity are: Use what you've got, improvise, make do. And for the true bricoleur the tools in the bag will have been selected over a long time by a process determined by more than pragmatic utility. These mental tools will be as well worn and comfortable as the physical tools of the travel-ing tinkerer; they will give a sense of the familiar, of being at ease with oneself; they will be what Illich calls "convivial" and I called "syntonic" in Mindstorms.

Papert also notes that there are different forms of syntonicity in the learning of mathematics, viz. ego-syntonic, body-syntonic and cultural syntonic which must all filter through a context for learning mathematics in which the aesthetic is fore-grounded. For Illich (1973: 17) the term convivial—"with life"—is intentionally and deliberately chosen to, "designate a modern society of responsibly limited tools" (p.6) that,

> designate[s] the opposite of industrial productivity… [but rather] autonomous and creative intercourse among persons, and the intercourse of persons with their environment…indi-vidual freedom realized in personal interdependence…the freedom to make things among which they can live, to give shape to them according to their own tastes, and to put them to use in caring for and about others.

While Papert connects syntonic appropriation with Illich's tools for conviviality he has not elaborated upon the connection. We find that there is a need to elaborate this connection more fully in our work as this we believe is part of where attention needs to be drawn to extend frameworks like M4T/MKT/TPCK. When Papert talks about syntonic appropriation he is referring to those felt processes by which an object or tool becomes "an object-to-think-with" or following Sfard's (2008) com-mognitive theory, an object to discourse with. Papert's discussion of syntonic appro-priation exists in the realm of Deweyan educative experience, i.e. of individual learning.

Illich, on the other hand, is very much concerned with the role of technology in society and the press that technology imposes on everyday life through increases in industrial productivity and efficiency. Illich's argument is the need for responsibly limiting tools such that the locus of control remains with the individual in serving a community, i.e. tool use is a way to bring to life the imagination of the tool user within an ethical space. The Logo programming language and its evolution in forms like Scratch or the EV3 Lego robotic visual programming language is one way in which tool use can be responsibly limited. We can also approach the idea of "respon-sibly limiting" through the concept of "enabling constraints" of complexity informed approaches (Davis and Simmt 2003). While the tools themselves provide some degree of responsibly limited active engagement and immediate feedback, the inten-tion of a designer-teacher-curator can powerfully focus and direct learners' atten-tion and consolidation of understanding through meaningful task design and ongoing dialogue. In this way the designer-teacher-curator participates (without over-determining the pace, unfolding and trajectory) in the process of the learners' appropriation of the tools as tools for conviviality.

Syntonic appropriation allows for (mental and physical) tools to become partners in intellectual and creative life, to become discursive tools that give life to individual learner's ideas and communities of practitioners and which thereby contribute to growth in understanding. This freedom to select and make things – intellectual independence – the space of learning and growth is an ethical space for both Papert and Illich and one might add an empathetic space for creating a life-giving or convivial community.

The mathematical tool/object-to-think-with that we are intending students to make a syntonic appropriation with is the number line. We worked to design/curate tasks that embed the number line as an object-to-think-with within students' initial appropriation of robots as objects-to-think-with. We intended that students might make a (more) syntonic appropriation with the number line in this context than with other presentations and previous experiences. That is to say, we intended for it to become a tool for conviviality in relation to learning mathematics. In the next section, we describe the tasks and our process that we designed / implemented / refined / curated for embedding the number line as an object-to-think-with.

Appropriable Challenges

A Curated Task for Learning Mathematics

In our previous work (see Francis et al. 2016), we investigated how enactivism was a good framework for studying children's engagement in spatial reasoning while programming robots to move. In this paper, we are using our understanding of spatial reasoning to work more explicitly to develop understandings of the rational/ decimal number line. We are putting more attention on the aspects of the task that enable this. We are curating our past experiences to direct students' attention and actions towards enabling their syntonic appropriation of the number line as an object-to-think-with (or tool for conviviality) as the specific domain over which their growth in mathematical understanding is observed.

In the following task, learners are invited to explore how the <move steering> programming block works to turn the robot. Figure 10.4 below is an example of the EV3 <move steering> programming block's steering set to 25. Our intention in this task is to intentionally vary the steering settings incrementally, and thereby draw students attention to specific observations about (1) how the wheels rotate, (2) how the robot travels in terms of the radius of the robot's turn and the circumference of its circular path, and (3) what the steering means in terms of differential percentage through direct questioning/ dialogue/ prompting.

A recording sheet functions to create a shared focus of attention for discussion. Note that collaboration and dialogue are intended. Designer-teacher-curator has intentionally designed the recording sheet in order to prompt certain awarenesses and questions (see http://stem-education.ca/files/SteeringRecordingsheet-oldEV3_2020.

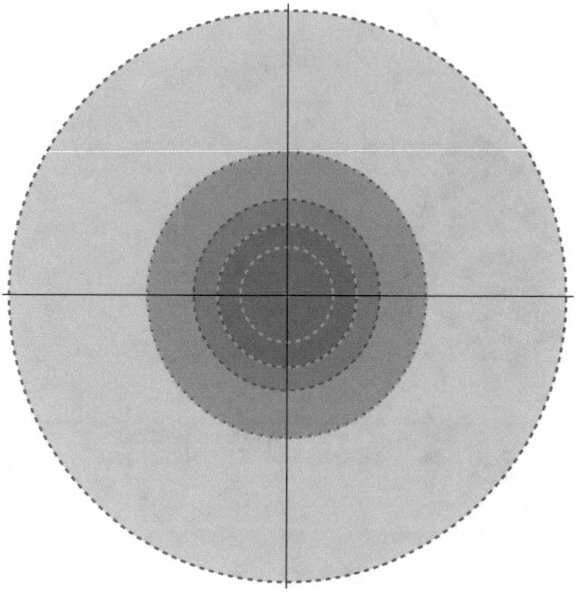

Fig. 10.3 Steering mat for move steering task (see http://stem-education.ca/files/SteeringMat.pdf)

pdf). In the first part of the recording sheet attention is drawn to how the wheels move and the direction of the robot's turn as the steering changes incrementally. For instance, when the steering is set to 25, then the right wheel rotates ½-wheel rotation forward and the left wheel rotates 1 rotation forward. The robot turns counter-clockwise.

Next, in order to have students attend to features of the robot's circular path, a mat was designed (see Fig. 10.3 above). Blue circles are the path of the outer wheel of the robot (at 25% steering the outer radius is 24 cm, at 50% it is 12 cm; 75% it is 8 cm and 100% it is 6 cm). The horizontal and vertical axes were included to give students access to the benchmark angles related to quarter, half and three-quarter turn and to allow for development of estimation strategies as well as serving as a marker for starting the robots off.

In order to easily follow the trace of the outer and inner wheels, the design of the mat allows students to discern that the robot has to be moved closer to the center of the circle as the steering increases. This reduces some of the cognitive load inherent in working with multiple aspects that vary simultaneously and allows a focusing on the critical learning intention of the task, viz. to learn how to turn the robot precisely.

Students are shown how to use the mat using the 25% steering – the robot is placed along one of the axes with the outer wheel on the largest blue circle (see Fig. 10.4 below). The number of wheel rotations is varied in order to get the robot to follow one complete circle. Only the number of wheel rotations is being varied at this time. This allows them to discern the radius and circumference of the robot's circular path with 25% steering as a (decimal) multiple of the number of wheel

Steering Set at 25%

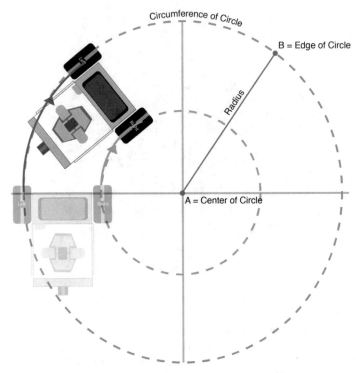

Fig. 10.4 A robot following a 24 cm radius circle with steering is set to 25

rotations. Next students are guided to investigate the radius and circumference of the robot's circular path for 50%, 75%, 100%, −75%, −50% and − 25% through the provided recording sheet.

Students are required to find which circular path the robot travels with incremental variations of the steering. For instance, when the steering is set to 25, then the left outer wheel of the robot travels around the circle with a 24 cm radius. Next they are asked to determine how many wheel rotations it takes to complete one circle (the circumference). The extension of the second part is to convert the number of wheel rotations to cm.

Lastly, in the final portion of the task, students are asked to pictorially model the first part of the task (the number of rotations for each wheel with incremental steering changes) using fraction bars (with the 'whole' being two fraction bars representing one complete wheel rotation forward and one complete rotation backwards. Then they are asked to convert the fractions to a percentage (which represents the steering on the <move steering> block). Figure 10.5, summarizes the details that the designer-teacher-curator wanted to call attention to.

	Steering is set to 0. The wheel rotation is set to 1. Both wheels rotate forward 1-wheel rotation. The robot moves straight. Differential is 0%.		Backwards / Forwards. Left, Right, Total
	Steering is set to 25. The left wheel rotates forward 1 rotation. The right wheel rotates forward $\frac{1}{2}$ rotations. Differential is $\frac{1}{4}$ or 25%.		Backwards / Forwards. Left, Right, Total
	Steering is set to 50. The left wheel rotates forward 1 rotation. The right wheel does not rotate. The robot pivots on the right wheel. Differential is $\frac{1}{2}$ or 50%.		Backwards / Forwards. Left, Right, Total
	Steering is set to 75. The left wheel rotates forward 1 rotation. The right wheel rotates backwards $\frac{1}{2}$ rotation. The robot turns tightly to the right. Differential is $\frac{3}{4}$ or 75%.		Backwards / Forwards. Left, Right, Total
	Steering set to 100. Left wheel rotates forward 1-wheel rotation. Right wheel rotates backwards 1-wheel rotation. The robot pirouettes right. Differential is $\frac{2}{2}$ or 100%.		Backwards / Forwards. Left, Right, Total

Fig. 10.5 Summary of how the robot turns with incremental changes to the steering (http://stem-education.ca/files/SteeringExplanationSummary.pdf)

In Fig. 10.5 above the steering parameter is being varied incrementally. It is the only critical aspect that is varied (everything else is held constant). This allows learners to discern the function of the steering block in terms of its gross effects on the robot's movement. However, to do this intentionally, the task has been designed and refined, or curated so that attention is explicitly drawn to the *number* (or *fraction*) of wheel rotations and the *direction* of the wheel rotation, and the *direction* that the robots turn through the provided recording sheet. Each of these is a dimension of variation that is opened up. Varying this one thing in a structured (incremental) way allows learners to notice/attend to changes in the robot's behaviour at a number of distinct levels—each of these is a new dimension of variation (critical aspect for further study). Learners can think in terms of wheel rotations only, or in terms of the direction of wheel rotations only or the direction of robot turning only, however, the next part of the task requires working simultaneously with all of these to develop an understanding of how the robot moves around a circle. In this part of the task students are working with mathematical ideas of percentages (though they are not necessarily aware of this at this point in the task), direction of motion, and rotation.

In the intentional design of this task we asked what growth in mathematical understanding is possible, i.e. for an individual learner what is it possible to learn? In terms of *primitive knowings* they know how to program the <move steering> block and download to their robot, they know magnitude of numbers, language for direction of movement (clockwise and anti-clockwise or forward and backward), how to move the robot straight precisely (they have done tasks to do that, how many wheel rotations to travel 100 cm), decimal numbers (involved in measurement), how to measure accurately with rulers.

The larger goal of the move steering task is to use mathematical modelling to understand the black box of what the steering means in terms of how the robot moves/turns. Within the larger task, attention is directed to enactively experience the number line as an *object-to-think-with* through the concepts of circumference and radius. In the next section, we describe a student's engagement in the move steering task.

Knowing Is Appropriating Modes of Doing

Examples Illustrating Theory

We remind readers that our argument is that a well-structured robotics inquiry (such as one like we described above, where the teacher provides the initial questions and overview of what needs to be done, and the students work independently to formulate and analyze findings – with support and guidance from the teacher) – can lead to what Pirie and Kieren (1994a, b) called growth in mathematical understanding (whose modes were exemplified in our anecdote in Fig. 10.1) by encouraging processes of syntonic appropriation of specific mathematical *objects-to-think-with* (the number line).

In this section we intend to illustrate how the Pirie-Kieren model can be used to highlight/draw attention to some of the growth in mathematical understanding within a curated robotics learning experience (Fig. 10.6 below).

Image making - After 3 attempts to make the 3rd turn, Luke positions the robot at the same corner of the pentagon.

Image making - Luke changes the last <move steering> programming block and downloads the program. He is reflecting on what he know the robot should do and trying to get the robot to follow the third corner.

Image making - Luke is aware that the robot must follow path, compares expectation (evident in their programming) with what actually happens. Robot curves to the right (is not using the steering at 100%).

Image making - She scrolls back to the beginning of the code, presses play and observes the robot She DOWNLOADS LUKE'S PROGRAM

Image having - Kara is aware of what the code should be. Next, she adds 1 new move steering block. She changes the steering settings. First, she sets it to 100%, then she sets it close to 50%. This will have the robot pivot on one wheel.

Image having - Kara is aware of what the code should be. Then she adds another move steering block. She changes the steering to 100. The robot makes the appropriate turn.

Fig. 10.6 Luke and Kara tracing a pentagon (See video https://vimeo.com/343271775)

In previous observations of children programming their robots to trace a polygon (Francis and Poscente 2017) we noticed that the children did not appear to move beyond the *image having* category of mathematical understanding. As Pirie and Kieren (1994b: 40) describe, the *image having* mode is characterised by a strong dependence on metaphor and working with metaphor thus, "mathematics *is* the image that they have and their working with that image." In the shift in understanding to property noticing "similie comes into play—"is" becomes "is like." In the context of the polygon task, a shift to *property noticing* would be when students notice the similarities between the triangle and other polygons. For instance, a triangle program consists of a straight-turn 3 times. Noticing that a square "is like" a triangle because it is also a collection of straight-turn but it is 4 times instead of 3 times.

In this chapter we created illustrations to call attention to exchanges between students, teachers and the technology. These illustrations (Figs. 10.1, 10.6, and 10.7) are excerpts of videos that were obtained during weekly robotics classes in a local school (weekly robotics classes were held and video recorded weekly

Fig. 10.7 Greg syntonically appropriating the unmarked number line the move steering task

throughout the year for the past 4 years). These particular exchanges were chosen after an exhaustive interpretive video interaction analysis (Knoblauch et al. 2013). We began an initial overview of relevant video selection. This initial overview required reviewing field notes for finding and selecting video for analysis. Next, the selected videos were reviewed and the selection was refined based on the quality of the images, sound, actions and interactions.

Transcripts of the video do not illustrate the actions of the participants. For that reason, sequenced still image freeze frames were extracted from the video. Sketches of the images were made to improve the comprehensibility of the verbalized text form and make it easier for the reader to understand the situation (Knoblauch et al. 2013). Consistent with McLeod (1990), the sketches were placed in a juxtaposed sequential comic strip format to convey interactions. This format is similar to studies by Plowman and Stephen (2008) and Heath et al. (2010). The speech bubbles represent the dialogue; the commentary represents the metaphors of number used. The removal of background information allows us to keep attention focused on speech, gesture, and actions only and removes distracting elements such as carpet, tables and chairs. We are not saying that the classroom context is not important and we are aware of the loss of other information such as around ethnicity and valid concerns about representation in research images, however, in this study and paper these are not our focus for analysis in looking at growth in mathematical understanding.

Developing Modes of Doing

Figure 10.6 is an illustration of an exchange between two Grade 4 students, Luke and Kara, as they attempted to trace a pentagon. The exchange occurred quite early in the year and the students had familiarity with making their robot move straight and turn. In this exchange, Luke is attempting to rectify an issue with the robot's movements. The program works for the first two straight-turn increments of the pentagon's path. They had success making the first two straight-turn segments of the polygon. But the robot veers off path for the third straight-turn. Luke has attempted to correct the last two blocks of the assembled code three times. Figure 10.6 begins with his fourth attempt.

As they tested and retested ideas, they engaged with multiple spatial reasoning elements simultaneously while moving back and forth between *image making* and *image having*. We did not observe the children move into *formalising*, nor were they able to program their robots' turns consistently. This pattern continued throughout the year on many other robotics tasks. Each corner or distance for the robot to travel encountered in future tasks was approached with a guess and check process. Not to negate the importance of working in the space of *image making* and *image having*, we wondered if robotics tasks and inquiries could be structured in manners that could elicit deeper mathematical understanding that translates into more precise robotics movements and turns. Hence we developed the move steering task (described previously).

Developing More Powerful Modes of Doing

In the graphic illustration (Fig. 10.7), Greg, aged 11, is working on a curated robotics task that is intended to help explain how the <move steering> programming block works. In this part of the task, Greg is identifying which circle the outer wheel of the robot travels along, the radius of that circle in cm, the circumference of that circle in wheel rotations, and the direction the robot travels for <move steering> set at 100%, 75%, 50%, 25%, 0, −25%, −50%, −75%, and − 100%. Figure 10.7 begins as Greg is attempting to determine the circumference of the outer circle that the robot follows when the steering is set to 25%.

As Pirie and Kieren (1994b: 43) note *formalising* is characterised by "a sense that one's mathematical methods work for all" relevant examples. Children who are formalising do not need the physical actions and images which brought them to the point of formalising." Greg knows which circle the radius traces for all the positive steering settings. From the previous week's tasks, Greg learned that negative steering is symmetrical to the positive steering, but the robot turns in a different direction. He applied this previous learning to complete the rest of the recording sheet [*formalising*] without needing to test each setting.

We have shown in the examples above that a well-structured (curated) robotics task can lead to growth in mathematical understanding. In this final section we still need to show that this occurs by encouraging processes of syntonic appropriation of specific mathematical *objects-to-think-with* (the number line). In the many years of working with young children and robotics we have observed a syntonic appropriation of the robot in achieving the overall challenge goals (get as close to the wall as fast as possible). However, what we now see as being enabled by the carefully curated robotics task is a narrower but very powerful syntonic appropriation of specific, relevant mathematical *objects-to-think-with* such as the number line. This we believe is close to Papert's (1980) intent and description with learning environments such as Logo and Illich's (1973) view of technology as a tool for conviviality. Syntonic appropriation is not a one-off event or experience, it occurs at differential temporal paces and at different conceptual and affective grain-sizes for individual learners.

In the second frame of Fig. 10.7 above, Greg holds the robot on the mat and physically moves the robot while closely observing how far the wheel rotates to get to the exact start place again. At this point the learner, the robot and the mat—specifically the outer circumference which represents an unmarked number line—are structurally coupled (in the enactivist sense) as one learning system. Greg is physically and concretely measuring the circumference with the wheel's rotations, similar to how a measuring wheel measures field sizes and so is coming to awareness of the need and existence of relevant sequential smaller units of decimal measure in relation to horizontal distance traveled. It is at such points of careful attention and focus that we believe processes of syntonic appropriation related to growth in mathematical understanding are at work.

In the follow up task in the subsequent week, every group was able to turn their robot precisely to complete a new challenge (to see videos of a turn: https://vimeo.com/415696291 and here https://vimeo.com/415697745). They knew it was a 100% steering turn. From our previous experience we have seen that other groups typically have approached turning on this task (and others) through a more guess-and-check approach and would take significantly longer to arrive at this understanding, some even after as much as a year of working with robots. Note, we are not claiming that these learners know everything about circumference and radius, but they have developed from very similar initial *primitive knowings* to more than an *image having* level of understanding about these concepts. We are claiming however that through the curated learning environment and enactive experiences learners are at the point of beginning to notice properties and formalise measurement with decimal numbers and are developing an enactive relationship with the rational number line (not the real continuum at this time).

Conclusion

The structure of a task matters especially for developing mathematical understanding. Through the examples in the paper we have shown our process of designing and curating a robotics learning task with the specific intention of directing learners' attention to the underlying number line as an *object-to-think-with*. The task affords students an opportunity to work with multiple instantiations of number – number as a count, number as a measure, number as distance moved/rolled and the existence of numbers between whole numbers or decimal/rational numbers – and grow their mathematical understanding of number. The structuring of the task, we have argued, encourages processes of syntonic appropriation such that learners have a personal and embodied meaning of the concept and the associated *object-to-think-with*.

When structurally coupled with the robot and the task learners are cognitively and affectively inserting themselves INTO (not merely onto) the number line in an enactive way, i.e. there is a meaning to being 'between' two points as a result not of discrete hops (as in whole number counting) but of continuous motion. When number lines are typically encountered in early elementary (if at all) they are used to model addition, subtraction and multiplication. These all involve discrete 'hops' or 'jumps' forward and backwards on the number line from one well-defined point to another. However, hopping, jumping, even stepping is only a small portion of the repertoire of movements made by human bodies, the majority of which are experienced (even if not consciously aware) as continuous movements. The specific task affordances allow students to put themselves into the robot in a way that allows them to 'experience' a continuous movement and so come to appreciate that there are a multitude of numbers between 2 and 3 in a meaningful way. It is in this way that the prior syntonic appropriation of the robot (building and earlier tasks) allows for the deeper syntonic appropriation of a powerful mathematical model – that of the (rational) number line and which leads to the deepening of multiple mathematical understandings (including that of number etc.) enabling both robot and number line to become a tool for conviviality in relation to learning mathematics. We acknowledge that there are many other aspects of mathematics that are likely being developed during and through this learning experience, however, we have chosen to focus on those aspects that we intentionally designed for, viz. appropriating the number line as an *object-to-think-with*.

Implications for Classroom Practice

Curated robotics experiences can provide opportunities for children to learn these concepts not from an external "objective" perspective (disassociated way) but from a dynamic embodied and enactive perspective in ways that are meaningful. These curated experiences through providing a sufficiency of structure enable individual and collective processes of syntonic appropriation and sense-making that serves to enable growth in mathematical understanding.

One of the implications we see can be framed as teaching is presenting appropriable challenges in contrast to presenting content or experiences alone. We note from our own experiences that we, as teachers, also experience growth in our own pedagogical understanding of the teaching of mathematics. Earlier versions of robotics tasks we have used were not as appropriable for discerning the underlying *object-to-think-with* of the number line. Students were doing similar things BUT their attention was not being directed. In this instance the intentional design of the mats and recording sheet together with the task focuses students' attention on how the Move Steering works to turn the robot precisely. Turning the robot precisely, or as intended, is associated with positive (perhaps joyful) affect. To be clear, we do not believe that the learners have yet accumulated/curated sufficient and diverse sets of experiences as yet, and complementary mathematics learning is still needed to help students to formalise their understandings.

Teachers also have a (legal) responsibility to the curriculum which entails finding and developing appropriable tasks that address mathematical concepts in ways that enable syntonic appropriation through responsibly limiting (and gradually and deliberately increasing) the set of conceptual tools or *enactive-objects-to-think-with* (in contrast to merely *mental-objects-to-think-with* which represent a terminal goal of learning from action). Our explicit goal was to have the learners make a syntonic appropriation of the number line as an *object-to-think-with* to serve learners future growth in mathematical understanding. This future growth involves connecting their learning of multiple mathematical concepts beyond number. A teacher's role is to connect this knowing of the *enactive-object-to-think* with other mathematical concepts. In our move steering task for example the circumference of a circle was first experienced as distance traveled. In the extension to this task (not presented in this paper), these experiences are shifted to more formalized understandings of radius and circumference with the algebraic expression of c = 2πr (where c = circumference, r = radius) as an analysis of data collected.

Ideally a teacher would be able to continue to direct students' attention and action in using the *object-to-think-with* with other mathematical topics and concepts. This is where the designer-teacher-curator's role emerges as one that exceeds that of each of the individual categories (designer or teacher or curator) as each set of skills and dispositions is necessary but insufficient on its own. This requires, we think, a collaboration perhaps amongst task designers, classroom teachers and teacher-researchers who knit their curated nets of knowledge and experiences together. We see this as important to develop resources and learning experiences that help focus or sharpen attention on the intended object or aspect of learning (Dehaene 2020; Marton 2014).

Implications for Theory/Research

Modes of Knowing: Connecting Enactivism and Embodied Cognition

One contribution we see this paper as making is explicitly identifying the periods of time during analysis for which enactivist frameworks and embodied cognition frameworks are useful in making sense of student learning. Earlier we noted that, with reference to Fig. 10.1, enactivism is useful for understanding learning within the environment while an embodied cognition approach would foreground the relationships between prior engagements and outcome. Thus, we reiterate that enactivism is concerned with the learning *in* action since it is the potential for action in the world that focuses attention and drives learning while embodied cognition is concerned with the learning *from* action and thus a later consolidation of enactive action. We encourage others to see if this particular juxtaposition and blending of theoretical frames is useful for advancing understanding of student learning across developmental time both *in* action and *from* action.

Modes of Doing: Syntonic Appropriation

Our second contribution is to work explicitly with Papert's ideas around syntonic appropriation which we see as a missing element in frameworks like MKT/M4T/ TPCK that acknowledge but do not deeply engage with the affective domain in teaching and learning mathematics. Specifically, we have introduced the idea of 'curating' experience as a useful metaphor in relation to the type of knowing that characterises the types of intentional learning spaces and opportunities that were designed and used in the task. Following Papert we investigated and reinforced his link with Illich's ideas of technology as a tool for conviviality through (initially) offering responsibly limited usage that thereby enables focused attention and growth in mathematical understanding.

Modes of Understanding: Pirie-Kieren Model

We have incorporated technology into the Pirie-Kieren model of growth in mathematical understanding. Technology can provide enactivist experiences of concepts that support and strengthen those inner modes (*image making, image having*, and *property noticing*). Our work supports their view of growth in mathematical understanding as a non-linear process. The Pirie-Kieren model we believe is mostly about interpreting learning. What we have done is demonstrate how technology (structurally coupled with humans and a carefully curated task) can be used to *influence* learning. In learning to use the number line as an *object-to-think-with* for fractional (decimal) numbers and the idea of a number as a measure/distance, the use of robotics technology was critical.

In the current learning environment in schools with multiple competing learning initiatives and increasingly constrained teacher time, robotics platforms like the EV3, a well curated task and close collaboration with teachers allowed for addressing these multiple learning presses (STEM, CT, Multiliteracies, etc.). As mathematics learning evolves to increasingly include and depend on technology, designer-teacher-curators will require a complex and complementary set of interdisciplinary skills to both interpret learning in situ and design occasions to meaningfully influence learning.

Appendix

Program of Studies	Achievement Indicators Document
9. Represent and describe decimals (tenths and hundredths), concretely, pictorially and symbolically.	Write the decimal for a given concrete or pictorial representation of part of a set, part of a region or part of a unit of measure. Represent a given decimal, using concrete materials or a pictorial representation. Explain the meaning of each digit in a given decimal with all digits the same. Represent a given decimal, using money values (dimes and pennies). Record a given money value, using decimals. Provide examples of everyday contexts in which tenths and hundredths are used. Model, using manipulatives or pictures, that a given tenth can be expressed as a hundredth; e.g., 0.9 is equivalent to 0.90, or 9 dimes is equivalent to 90 pennies.

Program of Studies	Achievement Indicators Document
10. Relate decimals to fractions and fractions to decimals (to hundredths).	Express, orally and in written form, a given fraction with a denominator of 10 or 100 as a decimal. Read decimals as fractions; e.g., 0.5 is zero and five tenths. Express, orally and in written form, a given decimal in fraction form. Express a given pictorial or concrete representation as a fraction or decimal; e.g., 15 shaded squares on a hundredth grid can be expressed as 0.15 or 15/100 Express, orally and in written form, the decimal equivalent for a given fraction; e.g., <u>50</u>/100 expressed as 0.50.

Demonstrate an understanding of addition and subtraction of decimals (limited to hundredths) by:
  Using personal strategies to determine sums and differences
  Estimating sums and differences
  Using mental mathematics strategies
	Predict sums and differences of decimals, using estimation strategies.
	Determine the sum or difference of two given decimal numbers, using a mental mathematics strategy, and explain the strategy.
	Refine personal strategies to increase their efficiency.
	Solve problems, including money problems, which involve addition and subtraction of decimals, limited to hundredths.
	Determine the approximate solution of a given problem not requiring an exact answer.

10. Compare and order decimals (to thousandths) by using:
  Benchmarks
  Place value
  Equivalent decimals.
	Order a given set of decimals by placing them on a number line that contains the benchmarks 0.0, 0.5 and 1.0.
	Order a given set of decimals including only tenths, using place value.
	Order a given set of decimals including only hundredths, using place value.
	Order a given set of decimals including only thousandths, using place value.
	Explain what is the same and what is different about 0.2, 0.20 and 0.200.
	Order a given set of decimals including tenths, hundredths and thousandths, using equivalent decimals; e.g., 0.92, 0.7, 0.9, 0.876, 0.925 in order is: 0.700, 0.876, 0.900, 0.920, 0.925.

References

Acosta, C. and Adamson, G. (2017). Glen Adamson: Curating attention. *True Living of Art and Design*

Alberta Education. (2016a). Mathematics Kindergarten to Grade 9 programs of study. In *Program of studies* (updated). Retrieved from https://education.alberta.ca/media/3115252/2016_k_to_9_math_pos.pdf

Alberta Education (2016b). Alberta K-9 Mathematics Achievement Indicators. Retrieved from https://open.alberta.ca/publications/9781460127292

Ball, D. L., Hill, H. C., and Bass, H. (2005). Knowing mathematics for teaching: Who know mathematics well enough to third teach and how can we decide? *American Educator*, 2005, 14–21, 43–46.

Braithwaite, D. W., and Siegler, R. S. (2018). Developmental changes in the whole number bias. *Developmental Science* 21 (2): 1–13.

Brown, L. and Coles, A. (2011). Developing expertise: How enactivism reframes mathematics teacher development. *ZDM—The International Journal on Mathematics Education* 43 (6–7): 861–873.

Buteau, C., Muller, E., Marshall, N., Sacristán, A. I., and Mgombelo, J. (2016). Undergraduate mathematics students appropriating programming as a tool for modelling, simulation, and visualization: A case study. *Digital Experiences in Mathematics Education* 2 (2): 142–166.

Davis, B. and Renert, M. (2014). *The math teachers know: Profound understanding of emergent mathematics*. New York: Routledge.

Davis, B. and Simmt, E. (2003). Understanding learning systems: Mathematics education and complexity science. *Journal for Research in Mathematics Education* 34: 137–167.

Davis, B., Francis, K., and Friesen, S. (2019). *STEM education by design: opening horizons of possibilities*. New York: Routledge

Dehaene, S. (2020). *How we learn: Why brains learn better than any machine...for now*. New York: Viking.

Di Jaegher, H. and Di Paulo, E. (2013). Enactivism is not interactionism. *Frontiers in Human Neuroscience* 6 (345): 1–2.

Francis, K. and Poscente, M. (2017). Building number sense with Lego robots. *Teaching Children Mathematics* 23: 310–312.

Francis, K., Khan, S., and Davis, B. (2016). Enactivism, spatial reasoning and coding. *Digital Experiences in Mathematics Education* 2: 1–20.

George, L. G. (2017). *Children's learning of the partititive quotient fraction sub-construct and the elaboration of the don't need boundary feature of the Pirie-Kieren theory*. University of Southampton, Faculty of Social, Human and Mathematical Science, Thesis

Grover, S. and Pea, R. (2013). Computational thinking in K-12: A review of the state of the field. *Educational Researcher* 42: 380–43.

Heath, C., Hindmarsh, J., and Luff, P. (2010). *Video in qualitative research*. London: Sage Publications

Hiebert, J. E. and Carpenter, T. P. (1992) Learning and teaching with understanding. In: D. A. Grouws (ed.), *Handbook of research on mathematics teaching and learning,* 65–97. New York: Macmillan.

Humphreys, G. W., Yoon, E. Y., Kumar, S., Lestou, V., Kitadono, K., and Roberts, K. L. (2010). The interaction of attention and action: from seeing action to acting on perception. *British Journal of Psychology* 101: 185–206.

Hutto, D. D. (2013). Psychology unified: from folk psychology to radical enactivism. *Review of General Psychology* 17: 174–178.

Illich, I. (1973). *Tools for conviviality*. New York: Harper and Row.

Keen, R., Carrico, R. L., Sylvia, M. R., and Berthier, N. E. (2003). How infants use perceptual information to guide action. *Developmental Science* 6: 221–231.

Khan, S., Francis, K., and Davis, B. (2015). Accumulation of experience in a vast number of cases: Enactivism as a fit framework for the study of spatial reasoning in mathematics education. *ZDM Mathematics Education* 47: 269–279.

Knoblauch, H., Tuma, R., and Schnettler, B. (2013). *Videography: Introduction to interpretive videoanalysis of social situations*. Frankfurt: Peter Lang.

Koehler, M. J., and Mishra, P. (2005). What happens when teachers design educational technology? The development of Technological Pedagogical Content Knowledge. *Journal of Educational Computing Research* 32: 131–152.

Lakoff, G. and Núñez, R. E. (2000). *Where mathematics comes from: How the embodied mind brings mathematics into being*. New York: Basic Books.

Marton, F. (2014). *Necessary conditions of Learning*. New York: Routledge.

Marton, F. (2018). Towards a pedagogical theory of learning. In: K. Matsushita (ed.), *Deep Active Learning*, 59–77. https://doi.org/10.1007/978-981-10-5660-4_4

McLeod, S. (1990). *Understanding comics: The invisible art*. New York: HarperPerennial.

Merleau-Ponty, M. (1978). *Phenomenology of perception*. London: Routledge.

Namukasa, I. (2019). Integrated curricular and computational thinking concepts. *Math and Code Zine* 4. https://researchideas.ca/mc/integrated-concepts/

National Academies of Sciences, Engineering, and Medicine (2018). *How people learn II: Learners, contexts, and cultures*. Washington: The National Academies Press.

Obersteiner, A., Dresler, T., Bieck, S. M., and Moeller, K. (2019). Understanding fractions: Integrating results from mathematics education, cognitive psychology, and neuroscience. In: A. Norton and M. Alibali (eds.), *Constructing number*, 135–162. New York: Springer.

Papert, S. (1980). Mindstorms: children, computers, and powerful ideas. http://worrydream.com/refs/Papert-Mindstorms1st ed.pdf

Papert, S. (1993). *The children's machine: Rethinking school in the age of the computer*. New York: Basic Books.

Pirie, S. and Kieren, T. (1994a). Growth in mathematical understanding: How can we characterise it and how can we represent it? *Educational Studies in Mathematics* 26: 165–190.

Pirie, S. E. B. and Kieren, T. E. (1994b). Beyond metaphor: Formalising in mathematical understanding within constructivist environments. *For the Learning of Mathematics* 14: 39–43.

Plowman, L., and Stephen, C. (2008). The big picture? Video and the representation of interaction. *Educational Research Journal* 34: 541–565.

Preziosi, D. (2019). Curatorship as Bildungsroman: Or, from Hamlet to Hjelmslev. In: M. Hansen, A. F. Henningsen, and A. Gregersen (eds.). *Curational challenges: Interdisciplinary perspectives on contemporary curating*. London, Routledge.

Rushton, S. K. (2008). Perceptually guided action: a step in the right direction. *Current Biology* 18 (1): R36–R37.

Seligman, M. P. (2011). *Flourish: A visionary new understanding of happiness and well-being*. New York: Simon and Schuster.

Sfard, A. (1991) On the dual nature of mathematical conceptions: Reflections on processes and objects as different sides of the same coin. *Educational Studies in Mathematics* 22: 1–36.

Sfard, A. (2008). *Thinking as communicating: Human development, the growth of discourses and mathematising*. Cambridge: Cambridge University Press.

Sierpinska, A. (1994). *Understanding in mathematics*. London: Falmer Press.

Simon, M. A. (2006) Key developmental understandings in mathematics: A direction for investigating and establishing learning goals. *Mathematical Thinking and Learning* 8: 359–371.

Skemp, R. R. (1976) Instrumental understanding and relational understanding. *Mathematics Teaching* 77: 20–26.

Varela, F. J., Thompson, E., and Rosch, E. (1991). *The embodied mind: Cognitive science and human experience*. Cambridge: MIT Press.

Wing, J. M. (2006). Computational thinking. *Communications of the ACM*, 49(3). https://www.cs.cmu.edu/~15110-s13/Wing06-ct.pdf

Chapter 11
Why Do Mathematicians Need Diagrams? Peirce's Existential Graphs and the Idea of Immanent Visuality

Vitaly Kiryushchenko

Introduction

The topic of this chapter is the relationship between mathematical reasoning, diagrams, and everyday visual experience. My goal here is not to discuss either the external causes of this relationship or the variety of ways in which mathematicians actually use pictures and diagrams in their work. There is ample literature on both topics, including accounts of how the links between numerical and spatial representations are rooted in the same patterns of brain activity (Gracia-Bafalluy and Noël 2008; Hubbard et al. 2005), discussions of particular ways in which conceptual material and images are combined in mathematical reasoning (Loeb 2012; Lowrie and Kay 2001; Martinec and Salway 2005; Pinto and Tall 2002) and studies of the cases in which the application of diagrammatic representation proves to be especially conducive to teaching math (Bakker and Hoffmann 2005, Boaler 2016, Danesi 2016, pp. 92–108, Hegarty and Kozhevnikov 1999, Kucian et al. 2011, Legg 2017, Prusak 2012). The question I would like to ask here is more general: Why at all do mathematicians need to use diagrams, images and other visualizations in their work?

One way to approach this question is to say that pictures and diagrams play in mathematical proofs the role of auxiliary tools (Hanna 2007; Mumma 2010; Brown 1999). According to this view, pictures and diagrams are used by mathematicians in order to facilitate their reasoning and then translate those pictures and diagrams into a formal calculus. Although the diagrams are constructed as elaborate staged observations that make certain steps of a mathematical proof visually available, from this perspective, they do not constitute an independent mathematical language and are but partial and imprecise models designed for the purposes of informal demonstration only (Barker-Plummer 1997; Kulpa 2009). Accordingly, on this view,

---

V. Kiryushchenko (✉)
York University, Toronto, Canada
e-mail: kiryushc@yorku.ca

© Springer Nature Switzerland AG 2020
S. A. Costa et al. (eds.), *Mathematics (Education) in the Information Age*,
Mathematics in Mind, https://doi.org/10.1007/978-3-030-59177-9_11

mathematicians do not actually build proofs directly on visual imagery, but rather use the latter to enhance the symbolic formalization of the former. This view does have a significant practical merit, as it proves suggestive of a variety of particular modes of use associated with diagrams in mathematics. Yet in spite of its practical merit, in terms of the question posed above, this view does not help much. From the more general perspective the question above represents, explaining the advantages of using something by saying that it is good for the purpose, is like explaining the effects of opium, as the doctor from Moliere's *Imaginary Invalid* famously puts it, by *virtus dormitiva*, or its *capacity* to do so. The immediate further questions, in this case, are "What is it exactly that makes diagrams, understood as tools, useful?" and "Why exactly are formal proofs not enough?"

Another answer to the same general question—the answer I am going to defend here—is to say that there is a tight relationship between the deductive character of mathematical reasoning and the very way mathematicians construct their diagrams. According to this view, all deductions, including mathematical ones, in order to be accomplished, require some sort of observation—and therefore, ipso facto involve visual experience. Charles S. Peirce, the principal proponent of this view, claimed that, although not all diagrammatic reasoning is mathematical in nature, there is no mathematical reasoning proper that is not diagrammatic (CP1: 54, CP2: 216, CP5: 148, where CP followed by volume number and then paragraph number refers to *Collected Papers of Charles S. Peirce* [Peirce 1931–1958]). Peirce also believed that, this being the case, mathematical diagrams could be construed not simply as partial supplementary aids to formal mathematical proofs, but as immediate visualizations of the deductive process as such. Peirce's view has two important consequences. The first consequence is that the very necessity of mathematical deductions should be considered internal to the diagrams mathematicians construct. The second consequence is that there has to be something about the very nature of ordinary visual experience that directly links the basic spatial relations supporting our visual integration, on the one hand, and our mathematical intuitions, on the one hand. Furthermore, Peirce was convinced that, if these two claims are a matter of fact, then it should be possible to construct a deductive mathematical language that would amount to a complete system of diagrammatic expression independent of formal symbolic proofs.

Visual Representation

In order to fully appreciate the possibility of the independent visual language Peirce had in mind, and see how the idea of such language might help us answer, in the Peircean vein, the general question formulated above, we will need to understand why exactly Peirce the mathematician attached so much importance to diagrammatic expression. Although, as it is commonly recognized, mathematical reasoning was the heartbeat that pumped blood through the veins of Peirce's entire philosophical system, Peirce did not have a full-fledged philosophy of mathematics. For the lack of a systematic, self-explanatory account, then, we first need to look for the sources of Peirce's interest in diagrams that are external to his philosophy.

All of those sources are well known, yet have never been considered together. Meanwhile, such consideration, however brief, might prove very helpful. First, Peirce confessed that he had a strong *personal* habit of thinking by means of visual images, and that he was inclined to attribute this capacity to his mathematical mind-set (MS 619: 8, 1909, where MS followed by manuscript number and then page number refers to the *Charles S. Peirce Papers*, Houghton Library, Harvard University). At the same time, visual experience, Peirce insisted, was at the core of ordinary linguistic competence. In one of the entries of his late diary, he makes a confession: "I do not think I ever reflect in words: I employ visual diagrams, firstly, because this way of thinking is my natural language of self-communion, and secondly, because I am convinced that it is the best system for the purpose" (MS 619: 8, 1909). As an educator, Peirce believed that it would be a good idea if some sort of diagrammatic logic were taught in schools prior to the grammar of any natural language (CP4: 619). In his correspondence over the years, Peirce confessed repeatedly that, to him personally, English was as foreign as any other tongue. Moreover, he linked his incapacity for linguistic expression to his left-handedness, which, as he explains one of his letters to a mathematician Cassius L. Keyser, in turn, framed his social interactions:

> But I am left-handed; and I often think that means that I do not use my brain in the way that the mass of men do, and that peculiarity betrays itself also in my ways of thinking. Hence, I have always labored under the misfortune of being thought "original." Upon a set subject, I am likely to write worse than any man of equal practice (quoted in Brent 1998: 43. As Brent (1998: 15) notes, Peirce in fact was able to use *both* of his hands in writing simultaneously. For example, he was able to shock his students by writing on the blackboard, ambidextrously and simultaneously, a logical or mathematical problem and its solution).

In an early draft of "A neglected argument for the reality of God" (1908), Peirce further clarifies the matter, stating that he was "accustomed to think of Reason and Authority as opposite ways of determining opinions, and to approve of the former alone" (MS 842, 180–181). According to one of Peirce's letters to his friend Victoria Welby, this attitude towards authority and social conventions in general might be partly explained by the fact that Peirce was "brought up with far too lose a rein," except that he "was forced to think hard and continuously" (Peirce 1958: 417). Another letter to Lady Welby contains a more extensive explanation that provides a useful overall link between Peirce's left-handedness, his troubles with written language, his disdain for conventionality and the meticulousness of his personal thinking habits:

> [A]s a boy I invented a language in which almost every letter of every word made a definite contribution to its signification. It involved a classification of all possible ideas; and I need not say that it was never completed…The grammar of my Language was, I need hardly say, modelled in a general way after the Latin Grammar as almost all ideas of grammar are to this day. It had, in particular, the Latin parts of speech; and it never dawned upon me that they could be other than they are in Latin. Since then I have bought Testaments in such languages as Zulu, Dakota, Hawaiian, Jagalu, Magyar…These studies have done much to broaden my ideas of language in general; but they have never made me a good writer,

because my habits of thinking are so different from those of the generality of people. Besides I am left-handed (in the literal sense) which implies a cerebral development and connections of parts of the brain so different from those of right-handed people that the sinister is almost sure to be misunderstood and live a stranger to his kind, if not a misanthrope. This has, I doubt not, had a good deal to do with my devotion to the science of logic. Yet probably my intellectual left-handedness has been serviceable to my studies in that science. It has caused me to be thorough in penetrating the thoughts of my predecessors,— not merely their ideas as they understood them, but the potencies that were in them (Hardwick 1977: 95–96).

As all these letters and notes, taken together, suggest, in Peirce's case, the importance of visual experience extends beyond the bounds of purely theoretical concerns and has some implications in terms of his personal intellectual habits. Visual thinking, personal difficulties in dealing with written language, the nature of logic in general (and the model of a universal language in particular), left-handedness, and the tendency to disregard conventions happen to be intimately connected with each other.

Peirce's preference for visual representations, of course, went beyond this knot of personal intellectual idiosyncrasies. From very early on in his career, both as a mathematician and as a philosopher, Peirce paid close attention to the role played in mathematical cognition by *maps*. As a mathematician, he was professionally involved in solving mathematical problems related to geological maps (CP7: 85), and proving the so-called "four color theorem" (CP2: 105 CP5: 490 NEM4: 216–222, where NEM refers to Peirce's *New Elements of Mathematics* [Peirce 1976]). He also developed a map projection known as the "quincuncial map," which represented a transformation of conformal stereographic projection and was one of the first maps created with an application of the theory of functions of a complex variable (Eisele 1963, Kiryushchenko 2012, Kiryushchenko 2015; W4: 68–71, where W followed by volume number and then page number refers to the *Chronological Edition* [Peirce 1982]). As a philosopher, Peirce considered diagrams in general as *maps of thought* (CP4: 530). He rejected the idea of likeness, or similarity as originating from the comparison of two simple, visually given qualities. Instead, he believed likeness to be the result of the application of a *mapping rule* describing a relation established between two sets, where a unique element of one set is *paired* with one single element of another set. Peirce's principal suggestion was that what underpins our perceptions of things as being *alike* is the isomorphism not of substances, but of *relations* (see also Stjernfelt 2007: 50–77, Paavola 2011). And he claimed that maps, together with geometric diagrams and algebraic equations, were the primary examples of such isomorphism (NEM4: xv, CP4: 530, Bradley 2004: 71–73). A mathematical function is routinely understood as a mapping relation between sets of numbers, which tells us how to go on with interpreting the dynamics of the function. According to Peirce, by analogy, a visual feature we perceive as common to, say, a portrait of a person and the person themselves, is a result of mapping one set of relations between facial features onto another, which reveals a character of the portrayed person based on the schematization of an anticipated facial change.

Diagrams

Another source of Peirce's interest in the role diagrams play in mathematical reasoning is the fact that Peirce, whose first degree was in chemistry from Lawrence Scientific School at Harvard, had a tendency to draw broad parallels between his graphical logic and the idea of chemical valence. In particular, he compared logical relations to chemical compounds. For instance, on his view, a *medad* (a relation whose arity is zero) is similar to a saturated chemical compound—such that may result, for instance, from joining two bonds of a bivalent radicle (CP3: 421), and a *dyad* is similar to one oxygen atom chemically bonded to two atoms of hydrogen, which constitutes a molecule of water, etc. The analogy Peirce drew between his logic of relations and chemical valences is well known and thoroughly studied (Parker 1998: 63–70, Roberts 1973: 17–25, Samway 1995). However, one historical aspect behind this analogy is rarely mentioned. Namely, it is that its source lies in the metamorphosis, which had taken place in chemistry in the mid-1840s, and which was triggered by the formulation of the chemical type theory.

The idea that the type theory and, later, the theory of valences brought about was that chemical compounds could be studied not as mixtures of actual substances, but as relational pictures, or diagrammatic schematizations of those substances. Chemists discovered that the relational structure of a molecule and transformations of chemical compounds could be *depicted* in a certain way, with the use of basic graphical conventions. Thus, it is the idea of chemical valences that actually gave birth to the first fully developed scientific language, which provided a diagrammatic projection of the (previously hidden) life of its natural object. This said, the reason why Peirce attached so much value to the analogy between his mathematical diagrammatic logic and the system of chemical valences is that he considered both logical and molecular graphs as messages capable of *saying what* the matter of fact is and, simultaneously, *showing how* it is to be interpreted. In both cases, seeing how the graphs develop into meaningful structures and understanding how this development works is one and the same process—or, better say, one and the same *act*.

Another source of inspiration for Peirce with respect to mathematical diagrams was his correspondence with Alfred Bray Kempe, a British mathematician best known for his proof of the four-colour theorem (later shown incorrect). In 1886, the Royal Society of London published Kempe's *Memoir on the theory of mathematical form*. In the opening paragraph of the *Memoir*, Kempe stated that his intention was to separate whatever is necessary for "exact or mathematical thought" from "the accidental clothing," as well as to offer an "exposition of fundamental principles" and "a description of some simple and uniform modes of putting the necessary matter in evidence" (Kempe 1886: 2). The fundamental principles of the mathematical thought, separated from the accidental geometrical, algebraic, and logical clothing, were presented by Kempe as a system of *diagrams* that consisted of spots connected by different types of lines. Kempe's diagrams were supposed to express the universal form of algebraic and geometrical representations that would reveal a deeper grammar of mathematical thinking common to both. Kempe sent Peirce a copy of the *Memoir*, and a few months later, in 1887, Peirce answered with some suggestions that caused Kempe to make revisions also published in the *Transactions of the Royal Society of London* later that year (W6: xlv).

Now that we have situated diagrammatic expression within Peirce's mathematical mindset and learned what areas of research beyond philosophy and pure mathematics conditioned Peirce's aptitude for visual thinking, we can see that all the interconnections between these areas and Peirce's personal intellectual idiosyncrasies are based on an amalgam of a few core ideas. These are the ideas of likeness as isomorphism, natural language, basic relational structure of things, and certain intellectual economy that prescribes us to pay attention to what is necessary while disregarding the accidental. As will transpire, what brings all these ideas together is the role that, according to Peirce, is played in mathematical reasoning by *observation*.

One of Peirce's entries for Mark Baldwin's *Dictionary of philosophy and psychology* (1901) reads as follows:

> In mathematical reasoning there is a sort of observation. For a geometrical diagram or array of algebraical symbols is constructed according to an abstractly stated precept, and between the parts of such diagram or array certain relations are observed to obtain, other than those which were expressed in the precept. These being abstractly stated, and being generalized, so as to apply to every diagram constructed according to the same precept, give the conclusion (CP2: 216).

Peirce further claims that, in any particular instance of mathematical reasoning (not only in the case of geometry, but also in the case of algebraic equations and syllogistic structures), "there must be something amounting to a diagram before the mind's eye," and that "the act of inference consists in observing a relation between parts of that diagram that had not entered into the design of its construction" (NEM4: 353, CP2: 279). *Inferring*, then, according to Peirce, is *observing* attentively what an experiment with a diagram brings about. To use one of Peirce's own examples, a particular case of Barbara syllogism, written down correctly, represents a simple diagram that clearly *shows* the relationship between the three terms involved, and, in doing so, actually *exhibits* the fact that the middle term of the syllogism occurs in both premises. Likewise, an algebraic equation is a rule that maps one relation between variables onto another in such a way that further manipulation could lead to the discovery of a series of new facts. Even a purely symbolic algebraic formalisation, then, is an *icon* that pictorially represents relations between the terms involved.

A simple geometrical example would be Pythagoras' theorem. The majority of the proofs of this theorem require that, in order to explain the relation among the three sides of a right triangle, a geometer should make a certain *rearrangement*. In the initial, Pythagoras's own version of the proof, it is the rearrangement of the four identical right triangles whose hypotenuses form a square. When describing the process of such rearrangement in some detail, Peirce adds that, in any other case similar to the two above, what we need is

> to set down, or to imagine, some individual and definite schema, or diagram—in geometry, a figure composed of lines with letters attached; in algebra an array of letters of which some are repeated. This schema is constructed so as to conform to a hypothesis set forth in general

terms in the thesis of the theorem. Pains are taken so to construct it that there would be something closely similar in every possible state of things to which the hypothetical description in the thesis would be applicable, and furthermore to construct it so that ..., although the reasoning is based upon the study of an *individual* schema, it is nevertheless *necessary*, that is, applicable to all possible cases (CP 4: 233; emphasis added).

In an unpublished work titled "Syllabus" (c. 1902), Peirce extrapolates this point about the link between manipulating images, deductive necessity and discovery of new truths to icons in general:

For a great distinguishing property of the icon is that by the direct observation of it other truths concerning its object can be discovered than those which suffice to determine its construction. Thus, by means of two photographs a map can be drawn, etc. Given a conventional or other general sign of an object, to deduce any other truth than that which it explicitly signifies, it is necessary, in all cases, to replace that sign by an icon. This capacity of revealing unexpected truth is precisely that wherein the utility of algebraical formulae consists, so that the iconic character is the prevailing one (CP2: 279).

A year later, in lecture VI of his Harvard *Lectures on pragmatism* (1903), Peirce goes as far as to claim:

All necessary reasoning without exception is diagrammatic. That is, we construct an icon of our hypothetical state of things and proceed to observe it. This observation leads us to suspect that something is true, which we may or may not be able to formulate with precision, and we proceed to inquire whether it is true or not. For this purpose, it is necessary to form a plan of investigation and this is the most difficult part of the whole operation. We not only have to select the features of the diagram which it will be pertinent to pay attention to, but it is also of great importance to *return again and again to certain features*. Otherwise, although our conclusions may be correct, they will not be the particular conclusions at which we are aiming (CP 5.162; emphasis added).

Based on these, as well as other, more complicated examples, Peirce further shows that it is never the case that, in solving a problem, simply thinking in general terms is enough. "It is necessary," he says, "that something should be *done*. In geometry, subsidiary lines are drawn. In algebra, permissible transformations are made. Thereupon, the faculty of observation is called into play. Some relation between the parts of the schema is remarked" (CP 4:23, Hull 2017: 149; Joswick 1988: 113).

As Peirce notes, any one of Euclid's theorems is first formulated in abstract terms. However, in the *Elements*, such abstract statement, from which only some trivial truths may be deduced, is followed by the construction of a geometrical figure, and then, upon observation, the initial statement is reformulated in new terms; this time—with reference to the figure constructed. This, in turn, is followed by modifying the figure (by moving certain parts of it, or adding new lines, or both), and ascertaining whether the modifications hold good relative to the second formulation. Once this is done, Peirce says, the words, "which had to be demonstrated," follow without any further restatement of the result in abstract terms. As he further notes,

[i]n like manner when we have finished a process of thinking, and come to the logical criticism of it, the first question we ask ourselves is "What did I conclude?" To that we answer

with some *form of words,* probably. Yet we had probably not been thinking in any such form—certainly not, if our thought amounted to anything...What the process of thinking may have been *has nothing to do with this question* (CP2: 55, emphasis added).

There is, in the end of all this construction and rearrangement, a moment at which the result is *shown* by the speediest way possible, and after which thought can only idle in creating trivial corollaries. Again, according to Peirce, geometry represents only one of many possible cases in which this ultimate point is revealed. In fact, in any logical process whatsoever, Peirce says,

[w]hen we contemplate the premiss, we mentally *perceive* that that being true the conclusion is true. ... Since the conclusion becomes certain, there is some state at which it becomes *directly* certain. Now this no symbol can show; for a symbol is an indirect sign depending on the association of ideas. Hence, *a sign directly exhibiting the mode of relation is required* (CP4: 75, emphasis added).

According to Peirce, mathematics can discover new regularities due to the following two features that diagrams exhibit. First, because there is always an array of possible transformations, which are implied by the very way a given diagram is constructed, and all of which will never be enacted. Second, because, due to the essential indeterminacy of perception, we cannot predict in advance what particular transformations out of the array will in fact be enacted, and what the ultimate result of those transformations will be (Stjernfelt 2007: 81–83). What these two features imply is that mathematics essentially is an observation-based *activity*, a habit-driven, and yet creative *practice* rather than a static deductive grammar that supplies rules for the contemplation of abstract mathematical forms (Campos 2009; Hull 2017). Within mathematical reasoning as a practice, visual imagination has a three-fold role to play. First, a mathematician forms a *skeletonized* iconic representation, a diagram, whether geometrical or algebraic, of the facts he is interested in considering. The principal purpose of the initial skeletonization, Peirce says, "is to strip the significant relations of all disguise," so that "only one kind of concrete clothing is permitted—namely, such as, whether from habit or from the constitution of the mind, has become so familiar that it decidedly aids in tracing the consequences of the hypothesis (CP3: 559). Second, a mathematician observes this diagrammatic picture until, at some point, "a hypothesis suggests itself that there is a certain relation between some of its parts." Third, he experiments upon the diagram in order to test his hypothesis, so that "it is *seen* that the conclusion is compelled to be true by the conditions of the construction of the diagram" (CP2: 278, CP3: 560, Joswick 1988: 108–109). Mathematicians, thus, use some basic features of spatial representation to construct skeletonized images, or diagrams, such that certain changes in the relations between parts of those diagrams and a further analysis thereof reveal the necessary deductive force of the argument the diagrams represent. Considered in this vein, the diagrams are not just illustrations of the reasoning process; they are the process itself, visualized. And the only *authority* that we have in this case is not symbolic conventions, but the reasoning process itself, immediately visually present.

Recall that mathematical diagrams show relations that are constitutive of their objects and that, at the same time, can be manipulated so that new truths about their objects are discovered. Although a diagram is constructed "according to an abstractly stated precept" (CP2: 216), not all possible relations between the parts of the diagram are initially predefined in the precept. In this respect, diagrammatic expressions, if sufficiently conventionalized, are like any other language in that the array of possible interpretations presupposed by their initial construction always exceeds the array of new interpretations available, given our current goals and our point of view. Some Peirce scholars (Ambrosio 2014: 257) extrapolate this link between iconicity and the generative aspect of language on *any* representation:

> The very process of constructing an icon matters for Peirce, as it reveals the very respects in which a particular sign stands for its object. What seems to emerge from Peirce's account is that the very relation of representation is itself the result of a process of discovery: 'constructing' an icon amounts to discovering, and selecting, relevant respects in which a representation captures salient features of the object it stands for.

Given this, it is not surprising that Peirce himself consistently links iconicity and language. In particular, he claims, for instance, that "in the syntax of every language there are *logical* icons of the kind that are aided by conventional rules" (CP2: 280; emphasis added). The suggestion here is that language is capable of conveying and storing information not only because it symbolically encodes this information and refers to appropriate external objects, but also due to the fact that its syntax iconically frames our perception. On this view, the very order of meaning to some extent depends on the visual schematisms set up by the general syntactic arrangement of a given language. On this view, in a sense, the way we put organize the symbols we use in writing reflects the way we *think*.

In using diagrams, what we have is, as it were, a system of keyholes, through which we see something only because we do not see all the rest. However, what is peculiar about the use of diagrams in *mathematics* is that, even though the possibilities are limitless, the mathematician is capable of anticipating changes between the parts of a given diagrams that are characterized by *necessity*. Peirce admits that the nature of this relationship between novelty and necessity presents an unresolved problem. He claims that, "how the mathematician can guess in advance what changes to make is a mystery" (NEM4: 215). However, one might speculate that this capacity has something to do with the interplay between two Peircean distinctions: the deductive force of mathematical reasoning vs. the compulsive force of perception, and the active power of the imagination vs. the passive receptivity of perception. Taken together, the distinctions constitute part of the reason observation, according to Peirce, is always involved in mathematical reasoning. And the link between the two, again, is provided by an analogy between *perceptual* and *mathematical* judgments:

> We speak of *hard facts*. We wish our knowledge to conform to hard facts. Now, the "hardness" of fact lies in the insistency of the percept, its entirely irrational insistency... But this factor is not confined to the percept. We can know nothing about the percept ... except through the perceptual *judgment*, and this likewise compels acceptance without any assignable reason. This indefensible compulsiveness of the perceptual judgment is precisely what

constitutes the cogency of mathematical demonstration. One may be surprised that I should pigeon-hole mathematical demonstration with things unreasonably compulsory. But it is the truth that the nodus of any mathematical proof consists precisely in a judgment in every respect similar to the perceptual judgment except only that instead of referring to a percept forced upon our perception, it refers to an imagination of our creation. There is no more why or wherefore about it than about the perceptual judgment, "This which is before my eyes looks yellow" (CP7: 659).

The analogy, as Peirce further describes it, is rather intricate and by no means self-evident. The receptivity of perception is passive, while the imagination is an active capacity. What Peirce is saying here is that perceptual content forced on the passive receptivity of perception and an imprint produced by my own active power of imagination share the same phenomenological quality. Just as a percept is forced upon our perceptive capacity, a mathematical truth is forced upon our imagination; there is no "why or wherefore" about either of the two. In the latter case, there is a parallelism between the internal imaginative experimentation with diagrams (the capacity to *predict* the dynamic pattern of future changes) and external visual perception based on the capacity to *adapt* to the ever-changing environment. This analogy between the diagrammatic mathematical visuality and our ordinary, everyday visual experience now finally needs to be clarified.

Existential Graphs

As has been argued above, according to Peirce, spatial imagination and abstract reasoning are involved in the process of manipulating diagrams not as two distinct mental faculties, but as two aspects of the same activity put to work together. The point is aptly summarized in Hull (2017: 147): "Peirce's conception of a diagram is fundamentally and inseparably both conceptual *and* spatial insofar as reasoning by diagrams engages the continuum of spatial extension in the reasoning process."

Mathematics, then, is a practice that makes use of a set of specific cognitive mechanisms in order to creatively schematize together the general and the particular, abstractions and images, thought and action. A mathematician is capable of conceptualizing and, *in doing so*, directly observing the world as an ordered variety of forms of relations, because in mathematical thought, visual integration and conceptual syntheses are as *mutually interdependent* as two sides of a sheet of paper. Naturally, this mutual interdependence of general concepts and individual images should have a medium. Therefore, there should exist the possibility for a diagrammatic language that, in using simple graphical conventions, would embody the unity of the visual and the conceptual, of the perceptual dynamics and logical inference. This, Peirce believed, should be a language capable of visually representing thinking as it happens—thinking *in actu* (CP4: 6).

The possibility of such language, introduced by Peirce in late 1890s as "Existential Graphs," is contingent on two facts. First, according to Peirce, any perception can only be of a change. Just as there is no feeling of one's skin when a feather is not drawn across it, there is no proper vision of an object without some perception of the corresponding stereotypical motion. A cheetah chasing its prey, a butterfly in flight, a pencil bouncing off the table—each of these moving images loads our perception with habitual expectations, without which the visual integration necessary for our grasping those objects as such, would not be complete. Accordingly, if motion of an object does not just tell us where an object is going, but helps us recognize the object as such, it shall do so whether this object is a an animal, a pencil, a mathematical function, or a thought itself. If mathematics, expressed in a system of diagrams, or graphs, is to borrow from the architecture of ordinary visual recognition, then, in order to capture thought in action, to grasp the *continuity* of thinking, what we need to work out is not just a set of graphical conventions, but also a corresponding set of moves. In short, we need a system of *moving* pictures in order to turn thought into a proper object of study.

The second fact is this. Peirce, I believe, would admit that visual perceptions are inferential. Simply seeing something as "red" requires the capacity to apply the concept "red." Besides, acquiring such a concept involves a long history of piecemeal adjustments and readjustments, gradually habitualized intakes and responses to various objects in various circumstances. And this requires mastering some inferential skills. Moreover, according to Peirce, there is no sharp line of demarcation between perceptual judgment and hypothetical (or abductive) inference. Both amount to an act of a fallible insight assembling different elements that were present in our minds before. In the case of abduction, "it is the idea of putting together what we had never before dreamed of putting together which flashes the new suggestion before our contemplation" (CP5: 181). Perceptual judgment, in turn, "is the result of a process...not sufficiently conscious to be controlled, or, to state it more truly, not controllable and therefore not fully conscious" (ibid.). Peirce's system of graphs represents a move in the opposite direction: with the help of simple graphic conventions, the graphs make inferences a matter of visual perception.

On the one hand, then, we have images supported by the inferential ties that hold together our linguistic competence. On the other hand, we have inferences encoded visually. There is thus an exchange between the external, inferentially informed imagery of ordinary perception, and the immanent, diagrammatic imagery of mathematical thought. In using a set of basic spatial intuitions, Peirce's graphs *show* how inferences work. Meanwhile, the cognitive mechanisms that allow us to make those intuitions into the moving objects that the graphs are, are the same as those that shape our ordinary perception. To put it slightly differently, manipulating the graphs, which leads to the discovery of new truths, is based on the same perceptual dynamics that characterizes ordinary vision. But the deductive force of a conclusion, which results from the manipulation, is revealed due to the visuality that is immanent to the mechanisms of inference.

Final Remark

To conclude, Peirce's graphs represent an intricate knot of relations between written language, ordinary visual experience, necessary mathematical reasoning, and imaginative experimentation. While an ordinary person is content with the passive exercise of external perception only, a mathematician makes a good use of the interplay between the external visuality of objects and the immanent visuality of inferences in order to combine the creativity of mathematical thinking and the robust deductive necessity of its results.

References

Ambrosio, C. (2014). Iconic representations and representative practices. *International Studies in the Philosophy of Science* 28: 255–275.

Bakker, A. and Hoffmann, M. (2005). Diagrammatic reasoning as the basis for developing concepts: A semiotic analysis of students' learning about statistical distribution. *Educational Studies in Mathematics* 60: 333–358.

Barker-Plummer, D. (1997). The role of diagrams in mathematical proofs. *Machine Graphics and Vision* 6: 25–56.

Boaler J. (2016). *Mathematical mindsets: Unleashing students' potential through creative math, inspiring messages and innovative teaching*. New York: Wiley.

Bradley, J. (2004). The generalization of the mathematical function: A speculative analysis. In: G. Debrock (ed.). *Process pragmatism: Essays of a quiet philosophical revolution*, 71–86. Amsterdam: Rodopi.

Brent, J. (1998). *Charles Sanders Peirce: A life*. Bloomington: Indiana University Press.

Brown J. (1999). *Philosophy of mathematics: An introduction to the world of proofs and pictures*. London: Routledge.

Campos, D. (2009). Imagination, Concentration, and Generalization: Peirce on the Reasoning Abilities of the Mathematician. *Transactions of the Charles S. Peirce Society* 45: 135–156.

Danesi, M. (2016). *Learning and teaching mathematics in the global village. Math education in the digital age*. New York: Springer.

Eisele C. (1963). Charles S. Peirce and the problem of map-projection. *Proceedings of the American Philosophical Society* 107: 299–307.

Gracia-Bafalluy M. and Noël, M. P. (2008). Does finger training increase young children's numerical performance? *Cortex* 44: 368–375.

Hanna, G. (2007). Visualization and proof: A brief survey of philosophical perspectives. *ZDM. International Journal on Mathematics Education* 39: 73–78.

Hardwick, C. (1977). *Semiotic and significs: The correspondence between Charles S. Peirce and Victoria Lady Welby*. Bloomington: Indiana University Press.

Hegarty, M and Kozhevnikov, M. (1999). Types of visual–spatial representations and mathematical problem solving. *Journal of Educational Psychology* 91: 684–689.

Hubbard E., Piazza M., Pinel P., and Dehaene S. (2005) Interactions between number and space in parietal cortex. *Nature Reviews Neuroscience* 6: 435–448.

Hull, K. (2017). The iconic Peirce: Geometry, spatial intuition, and visual imagination. In: K. Hull and R. Atkins (eds.), *Peirce on perception and reasoning: From icons to logic,* 147–173. New York: Routledge.

Joswick, H. (1988). Peirce's mathematical model of interpretation. *Transactions of the Charles S. Peirce Society* 24: 107–121.

Kempe, A. (1886). A memoir of the theory of mathematical form. *Philosophical Transactions of the Royal Society of London* 177: 1–70.

Kiryushchenko V. (2015). Maps, diagrams, and signs: Visual experience in Peirce's semiotics. *International Handbook of Semiotics*, 115–124. New York: Springer.

Kiryushchenko V. (2012). The visual and the virtual in theory, life and scientific practice: The case of Peirce's Quincuncial Map Projection. *Semiotic and Cognitive Science Essays on the Nature of Mathematics*, 61–70. Munich: Lincom.

Kucian K, Grond U., Rotzer S., Henzi B., Schönmann C. (2011) Mental number line training in children with developmental dyscalculia. *NeuroImage* 57: 782–795.

Kulpa, Z. (2009). Main problems of diagrammatic reasoning. Part I: The generalization problem. *Foundations of Science* 14: 75–96.

Legg, C. (2017). Diagrammatic teaching: The role of iconic signs in meaningful pedagogy. In: I. Semetsky (ed.), *Edusemiotics: A handbook*, 29–46. Singapore: Springer.

Loeb A. (2012). *Concepts and images*. Cambridge: Springer Science.

Lowrie, T. and Kay, R. (2001). Relationship between visual and nonvisual solution methods and difficulty in elementary mathematics. *The Journal of Educational Research* 94: 248–255.

Martinec, R. and Salway, A. (2005). A system for image-text relations in new (and old) media. *Visual Communication* 4: 337–371.

Mumma, J. (2010). Proofs, pictures, and Euclid. *Synthèse* 175: 255–287.

Paavola, S. (2011). Diagrams, iconicity, and abductive discovery. *Semiotica* 186: 297–314.

Parker, K. (1998). *The continuity of Peirce's thought*. Nashville: Vanderbilt University Press.

Peirce, C. S. (1931–1958). *Collected Papers of Charles Sanders Peirce*. Vols. 1–8. C. Hartshorne, P. Weiss, and A. Burks (eds.). Cambridge: Harvard University Press.

Peirce, C. S. (1958). *Values in a universe of chance. Selected writings*. P. Wiener (ed.). Mineola: Dover Publications.

Peirce, C. S. (1982). *Writings of Charles S. Peirce. A chronological edition*. Vols. 1–6. M. H. Fisch, E. Moore, C. Kloesel, and Peirce Edition Project (eds.). Bloomington: Indiana University Press.

Peirce, C. S. (1976). *The New Elements of Mathematics*. C. Eisele (Ed.). Vol. 4. The Hague, The Netherlands: Mouton.

Pinto, M. and Tall, D. (2002). Building formal mathematics on visual imagery: A case study and a theory. *For the Learning of Mathematics* 22: 2–10.

Prusak, N. (2012). From visual reasoning to logical necessity through argumentative design. *Educational Studies in Mathematics* 79: 19–40.

Roberts, D. (1973). *The Existential Graphs of Charles S. Peirce*. The Hague: Mouton.

Samway P. (ed.). (1995). *A thief of Peirce: The letters of Kenneth Laine Ketner and Walker Percy*. Jackson: University Press of Mississippi.

Stjernfelt, F. (2007). *Diagrammatology: An investigation on the borderlines of phenomenology, ontology, and semiotics*. Dordrecht: Springer.

Chapter 12
Procedural Steps, Conceptual Steps, and Critical Discernments: A Necessary Evolution of School Mathematics in the Information Age

Martina Metz and Brent Davis

Introduction

Early in 2010, the Organization for Economic Cooperation and Development (OECD) published an online document in which it distinguished among "formal," "informal," and "non-formal" education. Many elements of this typography were predictable. Of the three, for example, only formal learning was identified to involve certified teachers, accredited curricula, and institutionalized settings. But there were also some unexpected elements. In particular, not many educational leaders expected a prominent—but unexplained and unjustified—statistic asserting that 75–85% of one's learning is other than formal.

That sort of datum is hard to contest. In fact, it would seem reasonable to argue that it is grossly underestimated. While not fully explained in the report, one can infer that the number to indicate the portion of an average life not dominated by attending school. That is, it was intended to emphasize the importance of lifelong learning. If that was the purpose, the point is simultaneously important and trivial. And that is perhaps why some in the educational establishment viewed the statistic with suspicion, as a not-so-veiled move to diminish schooling's long-held authority in matters of defining, offering, and certifying learnings.

In this regard, the technological context of the OECD's pronouncement is significant. It was a statement on learning in the Information Age. With advancements in and ubiquity of communication and storage technologies, traditional schools can no longer maintain a pretense of guardians of and gatekeepers to cultural knowledge. While broad awareness of that pretense has not yet contributed to substantial transformation in the institution, it would seem reasonable to expect that formal learning—that is, schooling—is on the threshold of significant transformation. In

M. Metz · B. Davis (✉)
Werklund School of Education, University of Calgary, Calgary, AB, Canada
e-mail: metzm@ucalgary.ca; brent.davis@ucalgary.ca

© Springer Nature Switzerland AG 2020
S. A. Costa et al. (eds.), *Mathematics (Education) in the Information Age*,
Mathematics in Mind, https://doi.org/10.1007/978-3-030-59177-9_12

this chapter, we muse on possible meanings and consequences of that realization, specifically as it pertains to mathematics education.

The social and cultural conditions of this potential transformation are not without precedent. Indeed, there are striking social and technological parallels between current circumstances and the historical moment that saw the original invention of public schooling. At the risk of oversimplification, key motivations for the creation of mass formal education in the western world included a dramatic shift in access to craft and scientific knowledge (enabled by printing presses and postal services), an associated convulsion in knowledge production (enabled by the co-amplifying influences of research institutions and business), an exponential growth of wealth, and the creation of legal systems that gave new rights to the disadvantaged as it recognized the dangers of increasingly inequitable distributions of that new wealth.

The modern school was partly a response to and partly a contributor to these intertwined convulsions. Simultaneously controlling and enabling, mandatory mass education was imposed as much to protect children from an exploitative labor market as it was to equip them with the basic tools needed to contribute effectively to that market. From the start, these basics were identified as the abilities to decode written texts, transcribe simple dictations, and perform uncomplicated calculations—or, more colloquially, "readin', 'ritin', and 'rithmetic." That is, the word "basics" originally signalled *minimal necessary skills* for workers. It pointed to some disciplines, but it said nothing about those disciplines themselves. Unfortunately, as the school became an entrenched and integral aspect of modern culture, the original, context-sensitive meaning of basics was lost. Thus, as society evolved, the basics remained stubbornly resilient. This detail is especially evident in school mathematics where, in the popular arena, "basics" is now typically assumed to refer to adding, subtracting, multiplying, and dividing—that is, not as a set of needs fitted to a particular context at a particular time, but as a reference to an assumed-to-be-natural foundation to mathematics. Indeed, the phrase "learning the basics" is often treated synonymously to "learning simple arithmetic."

Consequently, the construct of basics has become an albatross around the neck of mathematics education. As we develop in this chapter, for example, the notion was as the epicentre of multiple twentieth-century "reform" efforts, which sought to replace the traditionalist emphasis on mastery of *procedural steps* with a focus on *conceptual steps*—that is, to reframe mathematical competence in terms of progress toward deep understanding rather than mastery of technical procedures. That shift was tethered to dramatic developments in psychology and philosophy that contributed to new understandings of learning, which in turn revealed that the beliefs that oriented the original design of public schooling are plainly indefensible. (Even so, they still prevail.)

Profound and consequential insights into learning continue to emerge, now driven principally by the cognitive sciences. In this chapter, we use the notion of *critical discernments* to draw together some provocative emerging ideas and to explore their educational relevance against the now-popular contrast of *procedural steps* and *conceptual steps*. In the process, we also attempt to interrupt the contemporary meaning of "basics" by illustrating our discussions with concepts that we assert are basic to this era, but that are currently given minor attention in most formal curricula. On that matter, we (the authors) are unaware of any mathematics curriculum revision or mathematics teaching reform effort within our lifetimes—anytime, anywhere—that has not been stymied by demands to attend to the "basics." Our hope is that giving heed to matters of basic to whom, when, and where might contribute to efforts toward change.

Companion considerations are the conceptions of "learning" and "teaching" that arose alongside and continue to function in symbiotic relationship with the original notion of basics. Our suspicion, rooted in decades of engaging with classroom teachers, curriculum developers, and policy makers, is that a reason that the tendency to conflate "basics" and "mathematics" is so pervasive and so resistant is that that the assumed relationship is part of a grander flock of associations—that is, of mutually confirming assumptions of the nature of knowledge, the processes of learning, and the mechanics of teaching. In that regard, it appears that efforts to conceive of a mathematics education that is fitted to the moment are complex: A revisionist conception must simultaneously address matters of appropriate content and defensible practice. That is, it must engage with three sets of questions, seeking to understand the conditions of learning (*who, when, where, why?*), to identify and situate content (*what?*), and to define classroom practices based on current knowledge of human cognition (*how?*).

We attempt to take on all three of these matters in this chapter, but in differentiated ways. We start by taking on the *how*—an entrance point that is more intended to uncover some of the intricate web of associations that have over recent decades hobbled intelligent and action-oriented engagements with the other two matters. After that reframing, we turn illustration-based engagements with the conditions and content questions, moving on the conviction that actual experience with a new form of mathematics pedagogy is likely to be more compelling than an academic argument.

Learning: From "Getting" to "Constructing" to "Differentiating"

"Learning" is one of those phenomena that is intimately familiar, but shallowly understood. This point is cogently illustrated through the website, Discourses on Learning in Education (https://learningdiscourses.com), which describes, contrasts, and clusters over nearly 900 (at the time of this writing) perspectives on—that is, metaphors for, definitions of, theories on, strategies of—learning that are represented in the current education literature.

One of those discourses is popularly known as "twenty-first-century learning" or "Deeper Learning"—which, as first hearing, might seem an obvious alignment with the themes of this chapter. A blend of several prominent contemporary discourses, Deeper Learning is explicitly concerned with transforming formal education in ways that fit with emergent personal, social, cultural, technological, and economic conditions. While there are many varieties of the discourse, they tend to cluster around the same set of educational goals (e.g., robust academic outcomes, higher-level thinking skill, positive attitudes, technological proficiency, honed social skills) and to be defined by a specific cluster of teaching strategies (e.g., centered on real-world issues, oriented toward problems that are relevant to learners, choice-rich tasks, access to diverse tools and resources, frequent formative assessments, flexible and frequent opportunities to collaborate).

On the surface, then, Deeper Learning sounds like a movement that is hard to critique. However, even a shallow examination of the discourse reveals that, in fact, it rests on pretty much the same assumptions on knowledge, learning, and education as the traditional, *shallower* approach to schooling that it is presumed to critique.

The learningdiscourses.com site was motivated by this sort of realization. It is designed to assist in making sense of and sorting through competing and complementary perspectives. The project is informed by contemporary research in the cognitive sciences, a transdisciplinary domain that brings together psychology, linguistics, computer science, neuroscience, anthropology, philosophy, and other realms of inquiry. The cognitive sciences focus on the tools and strategies used by humans to make sense of the world, including especially tactics employed to maintain illusions of certainty against a reality of gaping holes in information, frequent flaws in logic, inevitable errors in recall, and implicit prejudices in perception. "Metaphor" figures centrally in these discussions, as both a means and a focus of analysis.

That emphasis is grounded in the twentieth-century realization that human thought is mainly analogical/associative rather than logical/deductive. Much of cognitive science research is thus trained on how metaphoric associations across domains of experience can orient perception, prompt action, bias interpretation, and infuse justifications. That focus turns out to be useful to sort through current discussions in education. As mentioned, the Discourses on Learning in Education site reviews and relates more than 900 currently active perspectives. That number is daunting. Somewhat less daunting, however, is that the number of core metaphors used across these discourses is much smaller (certainly under 50), and fewer than a dozen have any significant traction. As well, major educational movements tend to be associated with specific metaphors.

For example, traditional education is strongly reliant on metaphors through which knowledge is characterized as some sort of stable *object*, by which learning comes to be understood as *getting* that object. The intertwined notions are evident in such phrases as "collections of facts," "gathering of information," "tossing around ideas," "picking things up," "holding a belief," "getting it," "getting to," and "learning stuff." Ancient in origin (see Ong 1982), the grounding knowledge-as-object metaphor can be taken to suggest that there is a real truth, out there, stable, eternal, independent of knowers, untainted, and benign. The cultural priority of these qualities was later amplified in the first scientific revolution, as the ideal of objective truth, and further amplified as a nascent global capitalism found ways to commodify knowledge, creating a marketable thing.

As a means to understand the manner in which humans experience their truths, the knowledge-as-object metaphor has its value. However, as a principle for structuring formal learning, it is lacking. Nevertheless, the cluster of associations that have arisen around this figurative notion has long served as a grounding principle in public education. When knowledge is understood as a set of objects, then it makes sense to conceive of curriculum development in terms of selecting the most-worthy objects and formatting encounters with them. It also makes sense to approach their study as a systematic mastery of their parts. It renders learning a matter of picking things up, packing them in, and bouncing them back. It enables the interpretation of intelligence as how much one can hold—and that highly troublesome notion undergirds a multi-billion-dollar industry focused on measuring these imagined capacities. Ultimately, an uncritical embrace of the knowledge-as-object metaphor defines both learner and teacher, the former as a vessel or recipient, and the latter as a conveyer or deliverer.

The poverty of this cluster of notions was a major focus of psychological research in the early-twentieth century. To bring the issue to the fore, researchers test-drove a variety of new metaphors for learning, with *associating* and *constructing* figuring most prominently. Efforts were made to conceive of learning as an iterative cycle of interpretation, by which one's knowledge was framed as a coherent-but-evolving web of associations. The associated rise of "constructivist" theories among educators in the last half of the twentieth century represented an attempt to format the conversation for educators. Around school mathematics, constructivisms served as the main theoretical engines in major reform efforts, as they were used to alert educators to the problems associated with entrenched-but-invisible object-based metaphors and to the possibilities of taking up action-based metaphors. Problem-solving, personal strategies, learning from errors, talk-aloud protocols, manipulative-based explorations, and a grabbag other constructivism-influenced emphases were soon taken up. One of the popular memes used to collect these new ideas, and to distinguish them with entrenched notions, was a distinction proposed by Skemp (1976) between procedural and conceptual. Procedural was used to tag traditionalist emphases on acquisition and mindless mastery, and conceptual signalled shifts toward construction and making meaning.

Unfortunately, the shift didn't have the impact that theorists hoped. Our suspicion is that the ease with which notions of "constructing" can be blended with notions of "acquiring" proved to be debilitating. That is, the proposed new cloud of associations was perhaps not sufficiently distinct, and so more often than not they were subsumed into established practices and structures. Indeed, even some leading mathematics education researchers seemed to miss the point. Consider Sfard's (1998: 5) conclusion as she contrasted object-based and action-based conceptions of knowledge and learning:

> Concepts are to be understood as basic units of knowledge that can be accumulated, gradually refined, and combined to form ever richer cognitive structures. The picture is not much different when we talk about the learner as a person who constructs meaning.

Further to the surrounding issues, perhaps no one should be surprised that efforts to reform mathematics curriculum have been no more impactful than efforts to reform mathematics teaching. The miscontrual of basics continues in force, a linearized trajectory through prespecified content still dominates.

Enter the cognitive sciences. In recent decades, some educators have started to move to a much more distinct set of metaphors that frame learning in terms of differentiating—a two-layered action of noticing aspects of one's experiences and noticing/construing associations among those noticings. This idea is part of the growing appreciation that humans aren't especially logical. In fact, sapiens are bad at deductive reasoning—and, left to their own devices, tend to fall back on situation-specific and opportunistic tactics to get through situations that would be better managed through systematic thought. Humans extrapolate from past events, they seek patterns in the moment, they impose familiar metaphors, they re-enact established scripts. Thankfully, humans have also learned to off-load the demands of logical thought onto mechanical tools—except, for some reason, in contexts such as most public schools, where there remains an insistence that learners attempt to master mechanical processes that no longer need to be mastered. (To be clear about the point here: We believe that, to learn mathematics, learners must master concepts. But, as we develop, that sort of mastery is quite distinct from the mastery of multi-step procedures.)

What's particularly interesting about the metaphor of learning as differentiating (i.e., noticing and knitting noticings) is its utility for revealing the intellectual poverty of so many educational practices. For example, an immediate consequence of taking this metaphor seriously is that one must be especially attentive to what, exactly, learners are supposed to differentiate, how to channel attentions, how to organize experiences to increase the likelihood of useful associations, and so on. That is, the notion of learning as differentiating takes us immediately to a different model of teaching—one that simultaneously reveals the incoherence of many contemporary educational obsessions while offering a frame for alternative attitudes toward teaching and curriculum. We develop this and associated ideas in the last half of this chapter. But, before getting there, we must take on our second question, on the nature of mathematics. How does the differentiating metaphor prompt us to look at mathematics, and what does mathematical knowledge tell us about how it should be learned?

Mathematics: From "Building" to "Structure" to "Network"

Through the history of modern education, mathematics teaching practice has been consistent with prevailing beliefs about the discipline. For instance, a prominent, and likely dominant, belief is that mathematics is like a building. It has foundations. It has levels, and those levels are ordered. Hence, teaching and curriculum should be attentive to establishing solid foundations and tracing out its levels in logical order. That is, not only is mathematics an object, it is a specific sort of object that dictates topics and orders. Variations on these themes have defined school mathematics since the 1600s, in no small part because they are so compatible with the knowledge-as-object and learning-as-getting metaphors.

The realization that assumptions about the discipline affords an alternative characterization of many efforts to reform school mathematics in the twentieth century. In particular, in the last half of the century, a large number of teachers and researchers who embraced a constructivist sensibility lined up behind a new definition of mathematics—namely, as *what mathematicians do*. Circularity of logic notwithstanding, this shift in definition meshed with constructivist principles of learner agency, inseparability of knower and known, and gradual unfolding of possibility. It also shone a light on appropriate teaching emphases. In that regard, mathematicians were seen to be principally focused on solving problems. Authentic problems. Real problems. Sometimes open-ended problems. This shift in emphasis tied in nicely with progressivist emphases on authenticity and relevance, among other foci. It also fit with emerging sentiments and sensibilities that were later to evolve into Deeper Learning, as described in the previous section.

The move also set up what came to be known as the "Math Wars"—an ongoing, mainly North American-based tension between, in simplest terms, believers in back-to-basics sorts and proponents of problem-solving. Since the late 1980s, the Math Wars have dominated discussions of math teaching practice. For our purposes here, the critical detail is not the explicit tension, but that the Math Wars are enabled and perpetuated by two incompatible sensibilities—that is, two grand flocks of implicit association, each of which is internally consistent, but neither of which is especially defensible. The first of these found its anchor in the assumption that mathematics is an object that exists independent of humanity. In this flock, math learning is about faithfully reconstituting a *fixed reality*. The second flock swirls around the conviction that mathematics is an *evolving structure*—a hallmark of human creativity that emerges when logic is a defining quality. In terms of pragmatic consequences, the structure metaphor grounded criticisms of linear curricula, overly parsed concepts, isolated skills, and procedural steps while it prompted attentions to rich problems, meaningful contexts, flexible sequencing, and conceptual steps.

Yet, somehow, most efforts to enact these notions have not gone well. Somewhat ironically, a likely reason for the failure was anticipated by the person most commonly associated with problem-based learning. Noticing the tendency of humans to frame differences in terms of polarities, more than a century ago John Dewey (1910) cautioned that seeing differences in terms of polar opposites might compel debaters to think that those opposites must bookend all possibilities. That assumption, Dewey (1910: 9) worried, constrained thinking rather than enabling it, as he concluded that, "in fact intellectual progress usually occurs through sheer abandonment of those questions together with both of the alternatives they assume…We don't solve them: we get over them."

For instance, it might be tempting to think that the full spectrum of possibility for school mathematics is captured between "traditional" and "reform" sensibilities. On the one hand, mathematics is seen as pre-determined and pristinely organized—that is, it is regarded as something *discovered*. On the other hand, mathematics is viewed as contingent and subject to human interest and whim—that is, something *created*. Surely the continuum defined by "something *discovered*" and "something *created*" encompasses everything.

In fact, however, almost everyone who has framed their thinking with the dyad of "something *discovered*" and "something *created*" has missed a blindingly obvious detail: both elements of the dyad assume a *something*. Both are indexed to an assumption that mathematics is a thing—and, not somewhat ironically, this detail shows up most powerfully around the notion of *discovery*. As mentioned, among Traditionalists, mathematics is usually seen as out there, in the real word, and therefore discovered. Once discovered, however, it makes sense to convey it. Among Reformists, mathematics is most often assumed to be created, but somehow that conviction leads to strong recommendations for discovery-oriented teaching—revealing that object-based assumptions on mathematics have not been jettisoned at all. Perhaps that is why, even though Reformists managed to awaken educators to learner agency by attending to what mathematicians do, the contents and outcomes of most mathematics curricula are scarcely discernible from pre-reform versions, even after a half-century of Reformist influence.

Getting over the Math Wars, then, may be dependent on a compelling and defensible alternative to the implicit-but-pervasive knowledge-as-object metaphor that continues to undergird almost all discussions of school mathematics. Fortuitously, many alternatives have been developed, especially over the past few decades. One that we find especially useful is the metaphor that *knowledge is systemic coherence across levels of organization*, from which it follows that learning might be associated with making and acting on differentiations that enable systemic coherence. That is, learning is about noticing and knitting noticings—and, in turn, that blend sets up a model of school mathematics that aligns with neither side of the Math Wars. And it doesn't land between them either.

Over the past few decades, in efforts to understand the nature of their discipline, many mathematics researchers have turned its tools onto the discipline itself. A consistent conclusion is that mathematics is a complex system (e.g., Foote 2007). Consequently, mathematics has a decentralized network structure. Other phenomena that have this structure include cultures, ecosystems, and brains. More pointedly, mathematics is *not* an object—and, with that, the premises of the Math Wars crumble. As do both Traditionalist and Reformist teaching.

So, how might an educator approach mathematics when knowledge is understood as systemic coherence across levels of organization? To answer that question, we focus not on the elements of mathematics but on how the elements of mathematics might be made available in learners' experiences. A decentralized network comprises both nodes and links—which, in the case of mathematics, have been associated with "principles" (i.e., stable aspects of existence, such as patterns and forms) and "logics" (i.e., different means of combining principles into systems of interpretation). Learning mathematics, then, is about differentiating—that is, becoming aware of principles (i.e., noticing) and applying logics (i.e., knitting noticings).

Correspondingly, teaching comes to be about channeling attentions and juxtaposing experiences to support appropriate linking.

And that takes us to a model of mathematics teaching informed by "Variation Theory," which we argue is fitted to this Information Age.

Why Variation Theory?

As the Math Wars have continued to polarize discussions regarding the best ways to teach math (Schoenfeld 2004), it is clear that we have not yet adequately answered Chazan and Ball's (1999) call to go "beyond being told not to tell." Marton's theory of variation (Marton and Booth 1997; Marton and Tsui 2004; Marton 2015) offers powerful insights that allow a clear alternative to both telling and discovering—and to the knowledge-as-object metaphors upon which they are based. While it is impossible to "transmit" understanding or "process" the products of perception, it *is* possible to offer deliberate contrasts that dramatically increase the likelihood that learners will perceive intended principles and relationships in a particular way. In addition to effective prompting techniques, this requires careful attention to both short and long-term structuring—which we describe with the contronym, *raveling*—of mathematical ideas to which we might prompt. Neither effective prompting nor long-term raveling feature prominently in discussions of traditional vs. reform approaches, which has likely contributed to the longevity of the Math Wars. Traditional methods *can* work. So can reform methods. But the alleged reasons they work (or reasons the other does not) may have more to do with elements of pedagogy that do not even enter the conversation; further complicating matters, success may be *mis*attributed to one or other "contemporary obsession" (Preciado-Babb et al. 2020).

In this part, we further develop the key ideas underlying variation theory and relate them to principles of variation pedagogy developed independently in China (Gu et al. 2004; Huang and Leung 2004; Lai and Murray 2014). Following that, below we offer an interpretation of variation theory that integrates Marton's theoretical principles and Chinese pedagogical principles into a nested set of variation *types* that we've found helpful for designing short and long-term pedagogical sequences that support fractal awareness consistent with the nature of mathematics.

Learning as Differentiation

Marton's theory is based on the premise of *learning as differentiation*, which might be contrasted with *learning as enrichment*; i.e., that perception is always necessarily partial, depending on what we separate from an undifferentiated whole. Emerging from this is a distinctive principle that lies at the heart of Marton's work: The *Principle of Difference* states that we discern new ideas when they are contrasted against a constant background. *Highlighting difference to prompt distinction-making* is itself clarified by contrasting it with the common practice of (attempting to) *highlight similarity to promote association-making*. While association-making is indeed important to learning, Marton argued that we do not discern *new* meanings by perceiving similarities among examples that otherwise differ—i.e., through induction. If we can't perceive something in one place, we won't see it in two, or three, or a hundred. We can, however, *generalize* similarity among *previously* discerned examples (we also make metaphorical associations, but we will take that up a bit later).

Distinguishing generalizing from induction is essential to understanding and effectively using variation. This can be tricky, because often the patterns of variation used to prompt generalization are the same ones we might be tempted to offer in the hope of prompting induction. But order matters: Once separated through contrast, ideas become perceptible and can *then* be generalized. The examples may be the same, but the manner in which associations are assumed to be made is not. What can be highlighted via contrast is also constrained by prior knowledge, but in a different manner: We can't simply prompt attention to advanced mathematical ideas unless the necessary contrasts have themselves been sufficiently prompted. Thus, mathematical ideas must be carefully "raveled" so that we prompt attention at a level where learners are able to make sense of offered contrasts. It turns out that the levels can themselves be usefully described in more general terms, which is an important elaboration of the variation types that we discuss below.

The Principle of Difference: Induction vs. Generalization

The Principle of Difference is both less intuitive and more powerful than often appears at first glance. When we are trying to explain something that is familiar to us, it can *seem* as though multiple examples *should* support deeper insight. This is likely because once something has been discerned, multiple examples *can* add clarity through expansion of the example space associated with that idea (Watson and Mason 2005, 2006). Again, however, this is about *generalizing* existing understanding. If the particular something hasn't yet been discerned, it's impossible to simply *induce* what the many examples are examples *of*: We can't *induce* meaning from similarity if a unifying feature is unavailable.

Mason, Burton, and Stacey (1982/2010) recommended "generalizing and specializing" as a way of seeking deeper insight; it is through exploring variations of a particular case that we often find insight into both *that* case and *a more general class of cases* to which it belongs. Importantly, however, this is not about *finding similarities among varied examples*, but about *finding the perturbations under which the broader category remains intact* (which may itself be influenced by the particular conditions of investigation).

In some cases, a single contrast can provide the necessary insight for generalizing a particular feature, which is why sometimes it's possible to "see the general in the specific" (Mason and Pimm 1984; Watson and Mason 2006): Doing so involves seeing certain parameters of a problem as *variable*, which provides the necessary contrast for generalizing. In summary: *separation* (through *contrast*, not induction) must precede *generalization*; further, the very same examples that are inadequate for separation are ideal for generalization. Once separated and generalized, different features may be simultaneously varied, or *fused*.

Separating, Generalizing, and Fusing

What exactly does it mean to separate through contrast? If we want to highlight the meaning of, for example, *apple*, highlighting difference would involve *contrasting* an apple with other things that are like apples in as many ways as possible but differ with respect to some essential feature. At the moment of discerning, both *apple* and the broader whole from which it has been separated (Food? Fruit? Spherical objects?) become namable, but these namable "things" are less important than the un-namable *difference* that defines them. In other words, the notion of difference is essential to transcending the knowledge-as-object metaphor that has proved so intransigent over decades of attempts to improve math education.

Once the notion of apple has been separated from a background—say of food, or fruit, or spherical objects (i.e., once *apple* becomes a discernible and thus name-able difference)—we may *then* generalize to a broader class of apples. Although this class may be *defined* in terms of what all apples have in common, it is generated and bounded through expansion of allowable differences (Can it still be an apple if it has pink flesh?). To generalize, we hold the notion of apple constant and consider the permissibility of particular variations of apple—which looks just like the pattern of induction mentioned above, except that we're using *difference* (not similarity) to test the limits of our definition of apple. In other words, generalizing is about per-ceiving *differences between differences* (i.e. variations of *apple*, which is itself dis-tinct from *non-apple*)—and thus could be considered the sort of level change that lies at the heart of the hierarchy we are proposing.

According to Marton (2015), new ideas must be prompted in a manner such that what is general and what is specific are discerned *simultaneously*. When distin-guishing apple from non-apple, it may be that a learner becomes aware of the cate-gory *fruit* of which apple is a particular type; Marton would call such a category a *critical aspect*. If apple is the first fruit to be so separated from the broader category, both *fruit* (as a category) and *apple* (as one member of that category) are perceived at the same time. In this case, the contrast highlights both the apple and a hierarchi-cal structure involving apples and fruit (i.e., both the critical *aspect* fruit and the critical *feature* apple). Thus, "What is general and what is specific are discerned simultaneously when a new meaning is appropriated. There cannot be any features experienced without the awareness of the aspect that unites them, nor can there be any aspect experienced without the awareness of features that belong to it. Differences and features that differ cannot exist without each other" (Marton 2015:

48). Once apple and fruit have been thus separated, we might be moved to lay out the particular criteria we see as essential to each. Having done so, we might hold up *new* examples to those criteria and thereby classify them as fruit, apple, or both. *Classification* is distinct from *generalizing* in that it sets particular features of a particular example against an articulated definition; generalizing, on the other hand, takes particular features of a particular example and identifies a space of possibility bounded by the constraints of the experiential context as opposed to by the defined criteria of a definition.

Watson and Mason (2005) helpfully referred to critical aspects in terms of *dimensions of possible variation* and critical *features* in terms of *range of permissible change*. To highlight allowable variations of apple, apple becomes the critical aspect, which can be generalized according to variation in (familiar) features such as color, shape, size, and flavor. In doing so, it is helpful to contrast and vary features one at a time: Apples can be various shades of yellow, green, or red. They range from quite round (Macintosh) to a bit lumpy (Red Delicious). They can be smaller than a tennis ball (crab-apple) or as big as a softball (Fuji). They can range from quite sour (Granny Smith) to very sweet (Fuji). Color, shape, size, and flavor are *critical features* (of the *critical aspect* apple), and each can be varied within certain parameters. Similarly, "yellow, green, and red" are features of the aspect color, but discerning color itself isn't the focus at this time; in the context of the apple, color is assumed as prior knowledge.

Even when critical features have been carefully *named* in an attempt to separate them for attention, teachers and resources frequently attempt to prompt attention to the *named thing* rather than to relevant *differences*—differences which might *then* be given a name. This is how easily Marton's induction insinuates itself into pedagogies where the metaphor of knowledge-as-object has not been interrupted.

To offer a simple example, "practice rounding" tends to involve practice sets that cluster varied numbers to be rounded to nearest ten, then numbers to be rounded to the nearest hundred, and so on. Direct contrast between rounding to the nearest ten and nearest hundred is thus not easily perceived. Alternatively, the pattern of variation in Fig. 12.1 offers *direct juxtaposition* of rounding to the nearest ten, hundred, thousand, and ten-thousand—and requires learners to round *the same number* to varied place values (note that it's important that students are prompted to work from top to bottom, as moving left to right offers the same pattern of variation we're hoping to interrupt). These contrasts offer meaningful information about the impact of rounding. *More* carefully chosen variation of *fewer* examples can be very powerful, because learners must practice *making key discernments* rather than merely practice

Round to nearest:	3421	5421	13421	16476
10				
100				
1000				
10000				

Round to nearest:	1329	1389	1789	6789
10				
100				
1000				
10000				

Round to nearest:	1119	1199	1999	9999
10				
100				
1000				
10000				

Fig. 12.1 Rounding

a procedure. The contrasts *between* selected numbers and between successive charts are also carefully chosen, but these differences between differences can only make sense if the first order of difference has been successfully prompted. If we treat the first level of difference as a *thing* rather than a difference, every level thereafter becomes inaccessible. In this sense, first-order differences become the essential criterion for what is truly *basic* to structurally coherent mathematics pedagogy.

Similarly, when learning to add multi-digit numbers, learners may be asked first to add without regrouping, then to add with regrouping. A key discernment in doing so, however, is recognition that tens in the ones' place *can* be re-grouped. In other words, re-grouping can only be perceived in contrast to *not* re-grouping (and vice versa). When examples and practice sets separate adding with and without re-grouping, such contrast is far less obvious. Alternatively, if *varied tens* in the ones' place are directly juxtaposed (and include *no* tens), particularly in a manner that highlights those tens, the meaning of re-grouping is more likely to be perceived see Fig. 12.2). At the very least, learners should have to decide whether or not to re-group (zero tens or one ten). In practice sets where every example involves trading a single ten, we have observed learners go through the entire set and placing a "1" in the box for trades. They are not *distinguishing* between trades and no trades (or between one ten and other numbers of tens); they are merely performing a step that is highly limited to the particular context of that practice set. Note that in the last example, the lack of color-coding, slight mixing of the ten pairs, and inclusion of an extra one introduce elements that widen the space in which learners are expected to make appropriate discernments. Depending on learners' background and confidence, these features might be introduced one at a time.

To summarize: Once an idea is *separated* through contrast, essential features may also be highlighted by varying them against a constant background and *generalized* by identifying *non-essential* features. They may also be *fused* by co-varying previously discerned (and possibly generalized) features. Returning to the apple,

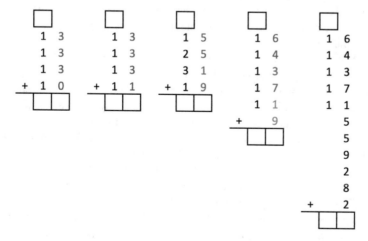

Fig. 12.2 Re-grouping tens

eventually, we recognize *apple* even when multiple features co-vary and even when those features co-vary in ways that produce novel situations. Even if we were suddenly confronted with a crab-apple-sized, Red Delicious-shaped, yellow-coloured apple, we would likely recognize it as an apple. Similarly, we can learn to add multi-digit numbers that involve re-grouping into *any* place value and with *any* number remaining in each place value. More significantly than what we can *do*, however, is that we may now *recognize* the very idea of re-grouping tens in a more broadly generalized sense that makes it available in a wide variety of other contexts, including all of the other so-called "basic" operations. Note that it's not the operations themselves that are basic, but the critical discernments upon which the traditional-defined basics need to be based.

While separating, generalizing, and fusing have to do with discerning the effects of multiple co-varying features, they do not fully account for how we generate mathematics or how we learn mathematics: Mathematical knowledge also expands both through *abduction* and through the *integration* of different ideas.

Abduction

If mathematical knowing has to do with organizing information into accurate hierarchies and identifying the *logical* implications implied by those structures, it may seem that mathematics is inherently a logical endeavour. But the *formation* of those hierarchies is an abductive—or more specifically, an analogical—process, which is not surprising considering that such hierarchies are created by analogical minds. Here, the similarity we decried in the context of induction assumes a prominent role, though difference is still essential to prompting new meaning. We generalize when we compare examples and decide whether identified differences are consistent with pre-specified criteria; we abduct when we transfer explanatory structure (consciously or unconsciously) between that which we perceive as similar. We *analogize* when we consider the appropriateness of transfer and do so (or refrain from doing so) intentionally.

To continue with our fruit example, it may be that a learner has already discerned apples and oranges but not considered their relationship. Doing so may simply involve combining them into a single category based on *specified* and *familiar* criteria; as noted about, this is *classification*. However, it may involve consideration of whether recognized shared features allow transfer of explanatory structure from one to the other. If so, we are talking not just about classification, but about *abduction*. Although difference is required to separate *new* ideas, the human mind is adept at perceiving similarities among previously discerned ideas. Abduction is *not* about seeing something *new*—it's about recognizing something familiar in a different context. From there, it is possible to consider whether what is known about each context may usefully or accurately be brought to bear on the other, though this aspect of abducting appears much less intuitive. Humans are deeply prone to jumping to conclusions based on perceived similarities. In any case, difference can only be perceived between ideas that have been previously discerned; i.e., differences can only be perceived between previously perceived differences. This would imply infinite regress if we assumed a blank-state infant mind, but we know that humans come into the world already attuned to particular differences from which all others eventually evolve (Lakoff and Núñez 2000).

Separating, generalizing, and fusing contribute to effective knowledge hierarchies. Marton (2015) addressed the importance of such hierarchies (particularly in the context of writing). Watson and Mason (2006) further emphasized the fractal nature of those hierarchies and highlighted the role of abstraction—which is consistent with what we're calling abduction—in their formation:

> However, to make mathematical progress the results of the images, models, and generalizations thus created have to become tools for more sophisticated mathematics. We see generalization as sensing the possible variation in a relationship, and abstraction as shifting from seeing relationships as specific to the situation, to seeing them as potential properties of similar situations. (Watson and Mason 2006: 94)

Taken together, the rejection of induction and the articulation of the role of difference in generalizing and abducting further support the importance of abandoning *mental things*. Doing so also helps resolve an apparent paradox highlighted by Watson (2017): If much of mathematics is defined in terms of *similarities* (defined by dependency relationships among variables) how can we use difference to prompt to similarity? One way of looking at this is that similarity is always between *similar differences*; if not, we couldn't have perceived them in the first place.

Integration

When used with well-raveled content, variation *theory* can offer powerful insights that contribute to effective variation *pedagogy*. To do so effectively, we must of course be clear about what we want to prompt attention to. But this is not as straightforward or intuitive as it might seem, particularly in the context of mathematics education. Marton (2015: 176) emphasized that "the object of learning that is used as a lens for inspecting the teaching may or may not be identical with the intended object of learning (i.e., the learning that the teacher had hoped to contribute to)." In mathematics it is frequently the case that instrumental learning is mistaken for relational learning (Skemp 1976). Carefully generated patterns of variation intended to teach particular mathematical ideas will ultimately fail if those ideas are defined only in instrumental terms; i.e., as *things* rather than *differences*.

Many who focus on step-based approaches emphasize that they *do* focus on the meaning of those steps. Even when the focus is on the underlying conceptual meaning of those steps, however, a procedure does not always offer an effective means of raveling the mathematical ideas required to make sense of that procedure. Many procedures require the integration of multiple ideas that, if not previously discerned and generalized, are very difficult to integrate (see Fig. 12.3).

Steps vs. Discernments

How does the standard algorithm for long division work?

Procedural Steps

1. Figure out how many times the divisor fits into the digit with the highest place value in the divisor. Write the number of times the divisor fits above the corresponding digit in the dividend (ignore remainder). This is the first digit in the quotient.

2. Multiply the first digit in the quotient by the divisor. Write the product beneath the first digit of the dividend.

3. Subtract.

4. Bring down the next digit of the dividend, and write it beside the number you got when subtracting. This will form a two-digit number.

5. Divide the two-digit number by the divisor, and write the answer you get beside the one you got in Step 1.

6. Multiply the second digit of the dividend by the divisor. Write the product beneath the two-digit number you divided in Step 4. Subtract.

7. Repeat Steps 2-4 until all digits have been divided (Divide, Multiply, Subtract, Bring Down, Repeat)

Conceptual Steps

1. Divide each place value one at a time. Starting with the highest place value, divide it into the number of groups specified by the divisor. Write the quotient above the digit you're dividing.

2. Multiply that answer by the divisor to find out how many of that place value have now been placed. Write the total beneath the corresponding place value in the dividend.

3. Subtract the number you got in #2 from the digit in the dividend that you were working with. The difference is the remainder that's left to be divided.

4. Combine the difference in #3 with the digit in the next place value; you can do this by simply bringing the next digit down. Unless Step 1 had a remainder of zero, you will now have a two-digit number.

5. Divide the two-digit number you found at the end of Step 3 by the divisor. Write the quotient beside your answer in #1.

6. Repeat Steps 2-4 until you have divided each place value. If there is still a remainder when you're finished, you can just state how many are left over.

Fig. 12.3 Procedural steps vs. conceptual steps (long division)

Steps vs. Discernments

How does the standard algorithm for long division work?

Conceptual Steps

1. Divide each place value one at a time. Starting with the highest place value, divide it into the number of groups specified by the divisor. Write the quotient above the digit you're dividing.

2. Multiply that answer by the divisor to find out how many of that place value have now been placed. Write the total beneath the corresponding place value in the dividend.

3. Subtract the number you got in #2 from the digit in the dividend that you were working with. The difference is the remainder that's left to be divided.

4. Combine the difference in #3 with the digit in the next place value; you can do this by simply bringing the next digit down. Unless Step 1 had a remainder of zero, you will now have a two-digit number.

5. Divide the two-digit number you found at the end of Step 3 by the divisor. Write the quotient beside your answer in #1.

6. Repeat Steps 2-4 until you have divided each place value. If there is still a remainder when you're finished, you can just state how many are left over.

Critical Discernments

CD#1: Division can be thought of in partitive or quotative terms. Here, we will focus on partitive division.

CD#2: When dividing, you can break the dividend into parts, divide each part into the number of groups specified by the divisor, then combine them (distributive property).

CD#3: If you break the dividend into parts that divide evenly by the divisor, only the final group (if any) will have a remainder. If you don't, the remainders from each group will also need to be combined. The combined remainders may be large enough to further divide.

CD#4: The standard algorithm for long division uses a special case of this strategy whereby the parts being divided are specified by the digits in each place value.

Fig. 12.4 Conceptual steps vs. critical discernments (division)

One of the obstacles that some seem to have with understanding critical discernments is that they think so long as steps are explained conceptually, they qualify as critical discernments. This ignores the importance of raveling: Often explaining a step in an algorithm involves multiple discernments (see Fig. 12.4), which is why many learners don't follow the conceptual explanation and beg to simply be given a list of steps. Critical discernments are raveled over time so that learners have made the necessary discernments that allow them to make sense of each new discernment.

Sometimes raveling algorithms involves integrating seemingly unrelated understandings that come together due to their role in the solution to a problem, but a well-raveled sequence should also have elements of progressive differentiation. In math, this often shows up in the form of seeing something as a special case of a broader principle. For example, the long division algorithm is a special case of separating-to-divide (CD #2 in Fig. 12.4), which itself is a refinement of the distributive property. Contrasting different ways of separating a number to divide opens "ways of separating" as a dimension of possible variation while simultaneously expanding the example space consisting of those ways.

Once again, generalizing and specializing emerge as two sides of the same coin. We generalize by varying and identifying boundaries for variation, not by looking for similarity among multiple examples. Nonetheless, it remains important to remain attentive to *implications* of such insights on both (or multiple) levels of the hierarchy and to how they might be elaborated; i.e., to what can be articulated in general terms and to the specific cases that comprise, limit, or extend articulated generalizations (Mason et al. 1982/2010).

This manner of viewing generalizing and specializing has implications for the current obsession with multiple strategies (Preciado-Babb et al. 2020). Rather than learning a variety of different ways to divide and then considering how they're *alike* (Marton's induction), we can progressively refine critical discernments about the nature of multiplication, division, and the distributive property. Each of the typical methods for division emerge from these critical discernments and are thus *already connected*. Again: *The general is recognized at the same moment that the particular is differentiated*; we see the general in the particular (Mason and Pimm 1984) *precisely when the particular emerges through differentiation*. Or, as Marton (2015: 48) put it: "There cannot be any new features experienced without the awareness of the aspect that unites them, nor can there be any aspect experienced without the awareness of features that belong to it. Differences and features that differ cannot exist without each other."

The distinction between procedural and conceptual steps is perhaps even more striking in the case of relating prime factors and factors (see Fig. 12.5). Offering a conceptual explanation of the procedural steps listed here would be grossly insufficient for most learners, because they would require a *sub-ravel* (and a sub-ravel of the sub-ravel) for each step before such an explanation could make sense.

Integrating multiple ideas is distinct from both fusion and from discerning dependency relationships among variables. It has to do with bringing diverse mathematical ideas to bear on a single context or problem, as in modeling and problem solving. Elsewhere, we have used a braiding metaphor (Preciado-Babb et al. 2020) to describe this difficulty: Learners need to braid the strands (i.e., each of the critical discernments on the right) before they can effectively attend to the rope itself. In other words, they would have to braid the strands at the same time that they're braiding the rope. This is also true of the discernments pertaining to long division, but the layers of sub-ravel requiring attention for the unknown to become perceptible may seem less daunting.

Steps vs. Discernments
How do prime factors determine number of factors?

Procedural Steps	Critical Discernments
1. Find the prime factors of the number.	CD#1: Every number can be written as the product of prime factors.
	CD#2: Every number has a *single* prime factorization.
2. Write the prime factors with exponential notation.	CD#3: Every number has a *unique* prime factorization.
	CD#4: Prime factors combine to make factors.
3. Add 1 to each exponent from Step 2. Multiply those numbers together to find the number of factors.	CD#5: The number of factors a number has is determined by the number of ways you can combine its prime factors.
	CD#6: If a number has one prime factor, repeated multiple times, the combinations are varying numbers of that factor.
	CD#7: If a number has two or more prime factors, clusters of one prime multiplied by clusters of the other(s) create *new* factors of that number.
	CD#8: The *number* of factors can be found by multiplying the number of possibilities for each prime factor in a number's prime factorization.

Fig. 12.5 How do prime factors determine the number of factors?

Conceptual and Procedural Variation

While there is value in talking about the theory of variation more generally, it is illuminating to consider how its principles are specifically implicated in the teaching of mathematics. In China, variation *pedagogy* particular to mathematics (Gu et al. 2004, 2017; Huang and Leung 2004; Lai and Murray 2014) was originally developed independently of Marton's theory, though collaboration among researchers from the two traditions has become common as researchers have recognized their complementarities. Nonetheless, there are some notable differences in emphasis (Huang et al. 2016; Pang et al. 2016; Watson 2017). In particular, Chinese variation pedagogy includes an explicit focus on *sameness* as well as *difference* (Gu et al. 2004). Watson (2017: 85) emphasized that much of mathematics takes *invariant* "dependency relationships" as its object of learning; in other words, similarity is, in fact, essential to mathematics. But earlier we insisted that humans are attuned to difference, not similarity, and that induction does not work. What's going on? We have found that the manner in which *sameness* is prompted is consistent with Marton's principle of difference, but it's important to consider more closely what it means to *use difference to prompt attention to similarity*—or more specifically, to the *underlying dependency relations that generate that similarity*. Prompting to relationship is much different than prompting to pattern (Hewitt 1992).

Gu et al. (2004) distinguished two types of variation important to teaching mathematics: (a) conceptual variation (CV) and (b) procedural variation (PV). Here, "conceptual" and "procedural" are used differently than is typical in Western contexts, where conceptual is roughly synonymous with Skemp's (1976) "relational," and "procedural" is roughly synonymous with his "instrumental." Both conceptual and procedural variation are about sense-making, and they are neither opposed nor competing. According to Gu, Huang, and Marton, conceptual variation offers examples, non-examples/counter-examples, and non-standard examples of a concept; thus, conceptual variation is roughly akin to Marton's contrast and generalization and to Watson and Mason's (2005, 2006) "example spaces." Conceptual variation offers the initial differences that bound a context for learning; it is concerned with clarifying and broadening the space of variation encompassed by a particular idea. Hewitt's (1999, 2001a, b) consideration of what is arbitrary and what is necessary in mathematics further highlights the importance of distinguishing what can and cannot be relationally defined and of teaching in ways that are consistent with this distinction. Even what is arbitrary, however, must be separated by prompting to *difference*; a particular definition or premise may be arbitrarily chosen, but once chosen, it cannot be arbitrarily prompted.

Procedural variation is further differentiated into three sub-categories: (a) varying the features of a problem (PV1); (b) comparing methods for solving a problem (PV2); and (c) considering how a single method can be applied to similar problems (PV3).

The procedural variations are collectively described as "progressively unfolding mathematics activities" (Gu et al. 2004: 319): "[P]rocedural variation intends to pave the way to help students establish the substantive connections between the new object of learning and the previous knowledge" (Gu et al. 2004: 340–341). In this way, successive examples may be experienced as "easier," but this is a particular kind of easier: They make it easier to make significant mathematical discernments, not just easier to complete a practice set or do a certain type of question. While some might see repetitive practice in a set of tasks designed with procedural variation in mind, it in fact involves deliberate change against a constant background; i.e., it's not the repetitive practice of a procedure but a way of highlighting relationships between particular mathematical variables. Lai and Murray (2014) argued convincingly that failure to distinguish between these two types of repetition likely lies at the heart of the perceived "Chinese paradox" (Huang and Leung 2004), whereby Western observers have sometimes struggled to make sense of how allegedly weak Chinese pedagogy consistently produces such strong results.

Differences based on logical hierarchy are also an important consideration when considering effective patterns of variation. A lesson (or a text) is experienced chronologically, but for learning to be effective, ideas offered within that chronology must be structured hierarchically; doing so involves prompting awareness of particular hierarchical relationships and how they are woven into increasingly dense and elaborated webs of association. Marton (2015) reported on several studies (outside of mathematics) where each new awareness was connected to a broader context. In such cases, learning was more effective than in cases where teaching was structured in linear sequence.

The different forms of variation can be seen in terms of a natural hierarchy with the potential to support the sort of meaningful long-term structuring of mathematics envisioned by Dienes (1960). Typically, however, *descriptions* of variation pedagogy often involve unrelated examples used to exemplify types of variation (Kullberg et al. 2017; Sun 2011; Wong 2008). While such studies are useful for prompting attention to the significance of fine-grained variation, it is not easy (a) to see the distinguishing features of different types of variation or (b) to see how they might be used to progressively elaborate an idea beyond the immediate context of a question set or a lesson. In China, coherent long-term raveling may be more clearly supported by carefully developed teaching resources, but elsewhere this is not always the case (Jianhua 2004; Bajaj 2013). Ma (1999) found that Chinese teachers, even those with less formal education than their counterparts elsewhere, showed more profound understanding of the elementary math they were teaching. In such cases, it is particularly vital that progressions highlighting hierarchical structure be embedded in quality resources that support both teachers and students in discerning complex webs of relationships.

Summary

We opened this part by noting that mathematics education has yet to articulate and put into broad practice a meaningful response to Chazan and Ball's (1999) observation that we need to offer teachers more than a directive not to "tell." Variation theory offers a way out of the apparent contradictions that emerge from many of the traditional vs. reform debates. It is not without potential pitfalls, however. Particular contradictions emerge when we attempt to use variation theory in conjunction with the knowledge-as-object. In such cases, variation theory tends to be distorted in one of two ways, depending whether the distortion occurs in a Traditionalist context or in a Reformist context.

Traditionalists may take variation as a means of offering gentler progressions and minimizing cognitive load. In other words, the subtle changes between questions are seen primarily in terms of gentle steps rather than in terms of meaningful contrasts deliberately chosen to make particular differences visible. While it is indeed important to attend to the limits of working memory, effective variation is about increasing clarity, not about making things easier. In fact, attempts to simplify by focusing on one thing at a time often result in the loss of the very contrasts necessary for effective prompting through variation.

Reformists working with the hope that learners will independently discover relevant knowledge-objects often fail to consider the hierarchy of differences required to make necessary contrasts perceptible in the first place. Open problems *can* offer spaces in which the variation of critical features assumes relevance (Runesson 2005), and learners can indeed take responsibility for generating their own patterns of variation emerging from those features (Watson and Mason 2005)—*if* they've discerned relevant dimensions of possible variation, *if* they have appropriate mathematical tools (developed through their own ravel) to manage that variation, and *if* they are not expected to weave the strands they are braiding, so to speak. Not all "paths" in the Reformists' journey follow a sequence that effectively supports attention to necessary differences, the fusion of multiple variables or the integration of prior knowledge.

When variation theory is paired with the metaphor of knowledge as a decentralized network, however, the importance of separating relevant features from an integrated whole and then re-integrating them in a manner that highlights an appropriate web of associations is much easier to talk about; i.e., the metaphor invokes both relationships and language that productively enable thinking and communicating about learning. Here, the significance of the particular ways in which mathematical ideas are raveled assumes prominence: Careful contrasts and wide spaces of variation typically open rich spaces of conceptual variation that subsequent procedural variations may continue to elaborate.

We have found it somewhat challenging to highlight how such hierarchies unfold in a longer-term ravel. To focus on the fine-grained distinctions significant at a particular level makes it harder to step back and focus on the relationships between levels in the hierarchy. To do so, we do not offer the same level of detail *within* each level that we might otherwise do, though we hope that the particular examples we've chosen sufficiently highlight the importance of fine-grained distinctions. Once again, the need to simultaneously attend to an intricate web of understanding at both the immediate and the long-term level fuels our insistence that a carefully developed resource is essential to the coherent, long-term elaboration of mathematical ideas.

Mathematics as Levels of Variation

We have found it helpful to conceptualize *types* of variation in terms of *levels* of variation defined by varied interactions among successive levels of difference (see Fig. 12.6). Level 1 separates and bounds key ideas with which we wish to work. This typically invokes what Hewitt (1999, 2001a, b) deemed the *arbitrary*. Levels 2–5 involve qualitatively different interactions among identified parts, each of which involves *necessary* implications (as opposed to arbitrary definitions) of the ideas established at Level 1. Levels 1–4 form a hierarchy of types: Level 1 uses

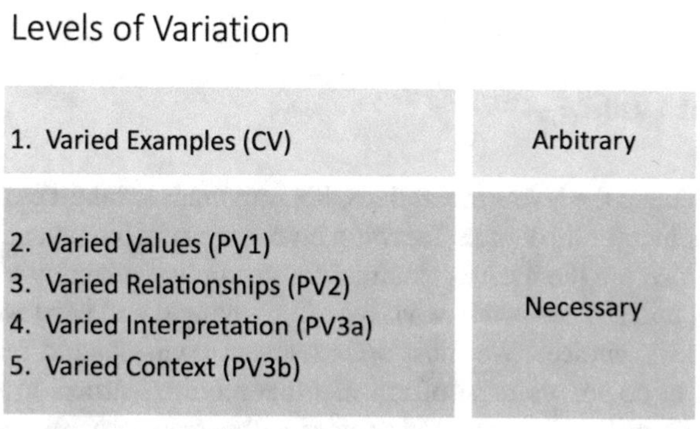

Levels of Variation

1. Varied Examples (CV)	Arbitrary
2. Varied Values (PV1) 3. Varied Relationships (PV2) 4. Varied Interpretation (PV3a) 5. Varied Context (PV3b)	Necessary

Fig. 12.6 Levels of variation

contrast to separate and generalize key features, Level 2 explores co-variation of those features, Level 3 contrasts relationships among different ways of co-varying (sometimes in the form of sequences and strategies), and Level 4 contrasts relationships among relationships. Level 5 focuses on interactions among previously identified features and relationships, including those that go beyond the object of learning identified in Levels 1–4.

Consistent with our claim that seeing knowledge as a decentralized network significantly influences how we make sense of teaching and learning, these levels might be helpfully compared with Bateson's articulation of logical types pertaining to the development of living things:

1. The parts of any member of Creatura [i.e., living things] are to be compared with other parts of the same individual to give first-order connections.
2. Crabs are to be compared with lobsters or men with horses to find similar relations between parts (i.e., to give second-order connections).
3. The *comparison* between crabs and lobsters is to be compared with the comparison *between* man and horse to provide third-order connections (Bateson 1979/2002: 10).

If we substitute ideas for organisms, we come very close to the framework we are attempting to describe. Hence we move from what Bateson referred to as *serial homology* (where each part within a particular organism is constrained during embryonic development by the previous parts) and *phylogenetic homology* (where new developments are constrained by shared evolutionary history) to what might be considered *parallel homology,* where parts do not act directly upon one another but may yet act in similar ways due to a history of evolving to meet similar evolutionary constraints. The three points above correspond to our Levels 2–4. Within and between each level, information is defined by *difference,* which is precisely why it can't be a *thing*. Difference exists in the space between. Each of the levels (1–5) are however, bounded—by what Bateson referred to as *context* and described in terms of a story, or pattern through time, that links varying elements in a space of shared relevance: "Any A is relevant to any B if both A and B are parts of components of the same 'story'" (p. 14).

Change and Choice

Before we articulate the levels themselves, it's important to take a closer look at the importance of change and choice. Teaching involves prompting attention to relevant differences, which are themselves essential for prompting attention to associations. Prompting has two key elements: what we offer (change) and what we invite learners to distinguish (choice). We must *offer* relevant contrasts and *invite* particular noticings, and to do so, we must offer tasks that require learners to *make relevant distinctions*. Attention to "change and choice" in this manner has a profound impact on what we call "practice:" Most of what gets called practice is "practice doing," but *good* practice is "practice discerning." In addition, examples and tasks must each use *appropriate* and *sufficient* contrast; i.e. they should be sufficiently different, uncluttered by distractors, and clearly juxtaposed in both time and space.

Finally, it is essential to acknowledge learners' roles in making themselves receptive to change: It is by moving our heads that our eyes receive differential signals regarding the position of objects in space and that we may thus perceive depth; this is also how our ears receive differential signals regarding the direction of sound, that we may thus perceive the direction from which a particular sound emanated; and it is by dragging our fingers over a surface that we may perceive differences that alert us to the shape or texture of whatever we are touching. Teachers may prompt to the significance of difference by inviting attention to relevant contrasts, but as learners come to *expect* such differences in the sequence of examples and tasks they are offered, they learn to do the mental equivalent of moving their head or dragging their finger, now over a *set of ideas*, but still with the aim of detecting the *differences* that contain relevant information. In the remainder of this section, we illustrate the five levels in a manner that we hope offers an abbreviated experience of relevant differences. A more elaborate sequence with many more opportunities for engagement may be found in Unit 3 of the Math Minds Online Course (Math Minds 2020).

Level 1: Bounding and Naming Differences

Level 1 defines the nature and limits of what we want to work with; i.e., to the particular aspects and features with which we wish to work. Level 1 is akin to conceptual variation in Gu et al.'s (2004) description of Chinese pedagogy and to separating in Marton's (2015) theory of variation.

The nature of Level 1 boundaries may be arbitrary in that there are infinitely many ways experience may be bounded, but they nonetheless define the premises from which necessary implications may be derived. It is here that the illusion of acquire-able "things" emerges. When we give differences a name, they appear as objects—as a "this" instead of a "this-as-opposed-to-that" or the conceptual "space-between-this-and-that". But if we lose sight of the difference that the name points to, we put ourselves and our students in a position from which no further insight is possible.

We take as our starting point exponents *as a special case of repeated multiplication in which all multiplicands are the same*. Prior experience with varied interpretations of multiplication will be assumed (see Fig. 12.7) as the starting point from which further differentiation may be prompted.

The *new* features to be developed through the full Level 1–5 sequence are (a) the extension from three multiplicands to an infinite number of addends, which has different implications for different interpretations of multiplication (Davis 2015), but which we wish to generalize to numerical laws, (b) the repetitive nature of multiplication when working with exponents, and (c) the mathematical notation used to

How is 2^3 like/unlike $2 \times 3 \times 5$?

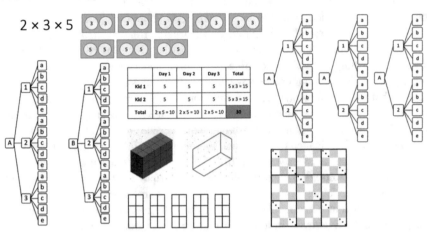

Fig. 12.7 How is 2^3 like/unlike $2 \times 3 \times 5$?

Multiplicative Repeats vs. Additive Repeats
(Level 1 Variation)

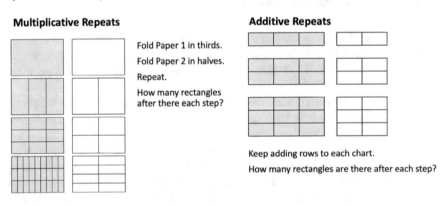

Fig. 12.8 Level 1 variation: What is repeated multiplication?

describe varied configurations of repeated multiplication. The decision to bound the multiplicands in such a way that they all match and the use of exponential notation to describe the possibilities that emerge in that space is arbitrary (Hewitt 1999), but clear contrasts are useful in prompting attention to these boundaries (see Figs. 12.8 and 12.9). We define what exponents *are* through contrast through what they are *not*, then generalize to less standard or more complex cases. Level 1 can often be characterized in terms of "yes-no-also," as in Fig. 12.9.

What **Are** Exponents?
(Level 1 Variation)

Exponents define *number of multiplicative repeats (starting at 1).*

YES	NO	ALSO: $(?)^3 = ? \times ? \times ?$
$2^0 = 1$	$2^5 \neq 2 \times 5$	
$2^1 = 2$	$25 \neq 10$	$(6 + 2)^3 = (6 + 2) \times (6 + 2) \times (6 + 2) = 8 \times 8 \times 8 = 512$
$2^3 = 2 \times 2 \times 2 = 8$		$(6 - 2)^3 = (6 - 2) \times (6 - 2) \times (6 - 2) = 4 \times 4 \times 4 = 64$
$2^5 = 2 \times 2 \times 2 \times 2 \times 2 = 32$	$10^5 \neq 10 \times 5$	$(6 \times 2)^3 = (6 \times 2) \times (6 \times 2) \times (6 \times 2) = 12 \times 12 \times 12 = 1728$
	$10{,}000 \neq 50$	$(6 \div 2)^3 = (6 \div 2) \times (6 \div 2) \times (6 \div 2) = 3 \times 3 \times 3 = 27$
$10^0 = 1$		$((6 + 2)^2)^3$ $((6 \div 2)^3)^2$
$10^1 = 10$	$2^5 \neq 5^2$	$= (6 + 2)^2 \times (6 + 2)^2 \times (6 + 2)^2$ $= (6 \div 2)^3 \times (6 \div 2)^3$
$10^2 = 10 \times 10 = 100$	$32 \neq 25$	$= 3^2 \times 3^2 \times 3^2$ $= 3^3 \times 3^3$
$10^5 = 10 \times 10 \times 10 \times 10 \times 10 = 10{,}000$		$= 9 \times 9 \times 9$ $= 27 \times 27$
		$= 729$ $= 729$

Fig. 12.9 Level 1 variation: What are exponents?

Varying Problem Features
Level 2 Variation: *Which column prompts* more effectively?*

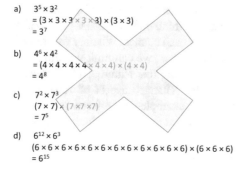

How Alike? (Induction)
What do all have in common?

a) $3^5 \times 3^2$
    $= (3 \times 3 \times 3 \times 3 \times 3) \times (3 \times 3)$
    $= 3^7$

b) $4^6 \times 4^2$
    $= (4 \times 4 \times 4 \times 4 \times 4 \times 4) \times (4 \times 4)$
    $= 4^8$

c) $7^2 \times 7^3$
    $(7 \times 7) \times (7 \times 7 \times 7)$
    $= 7^5$

d) $6^{12} \times 6^3$
    $(6 \times 6 \times 6 \times 6 \times 6 \times 6 \times 6 \times 6 \times 6 \times 6 \times 6 \times 6) \times (6 \times 6 \times 6)$
    $= 6^{15}$

How Different? (Generalization)
What happens if I change...?

a) $3^4 \times 3^3$
    $(3 \times 3 \times 3 \times 3) \times (3 \times 3 \times 3)$
    $= 3^7$

b) $8^4 \times 8^3$
    $(8 \times 8 \times 8 \times 8) \times (8 \times 8 \times 8)$
    $= 8^7$

c) $8^4 \times 8^5$
    $(8 \times 8 \times 8 \times 8) \times (8 \times 8 \times 8 \times 8 \times 8)$
    $= 8^9$

d) $8^4 \times 9^5$
    $(8 \times 8 \times 8 \times 8) \times (9 \times 9 \times 9 \times 9 \times 9)$
    $= 8^4 \times 9^5$ (can't be combined)

*Note this is just the first part of the prompt—learners would then *engage in a similar set of practice prompts.*

Fig. 12.10 Level 2 variation: Varying base, exponent

Level 2: Varying Features

Having thus defined the space in which we wish to work, we may now focus on variation of those apparent "things" or features. This puts us at Level 2, differences between differences, which has much in common with Marton's generalization and with the first type of procedural variations (PV1) described by Lai and Murray (2014); i.e., variation of problem conditions. This is highlighted in Fig. 12.10, which also attempts to clarify the difference between an inductive approach to varying problem conditions (not recommended) and an approach focused on difference (recommended).

Already in Fig. 12.9, variations of "yes" and "no" were offered, but the primary purpose of given contrasts was to identify relevant features (base, exponent) and to highlight the rules for interpreting exponential notation. At Level 2, we start to explore the *implications* of varying features identified at Level 1. In short, from Level 1 to Level 2, the goal shifts from defining boundaries to exploring implications of change. Importantly, changing one feature (A) has a resultant change on another (B), and the focus of attention shifts to this *relationship*, which might be seen as a sort of *difference between A and B*, and which we will call C. It's important that first one feature varies, then another, and then both together—Marton's *fusion*. Changing which feature is known and which is unknown can deepen understanding of the relationship between identified variables.

On the left side of Fig. 12.10, several examples are given in which learners must add exponents to get an answer. However, each question uses new bases and new exponents, which can make it harder to see the impact of change. In such cases, it is easy to fall into the trap of the teacher asking learners to "guess what is in my mind" (Mason 2010). While what is the same may seem obvious to the teacher, there are, in fact, a variety of features that are the same, and it's not always easy for learners to zone in on the intended one. In cases like this, teachers typically end up *giving* the rule, then ask learners to apply it in multiple cases. In so doing, it essentially gets reduced to a procedure rather than a generalized relationship.

The examples on the right are also cases where learners must add exponents, but now only one feature changes in each question, which makes it easier both to identify the intended feature *and* to see the *impact* of each change. Whether we are varying *within a particular law* or *between multiple laws*, we can limit variation to all but one feature, then observe the impact of changing that feature. Note that examples we've highlighted primarily focus on the "change" aspect of "change and choice." Through careful questioning as varied examples are offered and through appropriate follow-up tasks, it is also important that the teacher require learners to *make* relevant distinctions.

Level 3: Varying Relationships

If relationships *within a particular exponent law* are the focus, then it makes sense to vary features of that law—one feature at a time—and to observe the effect of doing so. By highlighting those changes—and their impact—the focus of attention shifts to relationships. However, distinguishing *among different exponent laws* should also become a focus fairly quickly (see Fig. 12.11). To do so, it is helpful to contrast the laws themselves while holding as many features constant as possible. In the set on the right, the bases and exponents are kept constant, while the operations change. Again, instead of doing several examples that all require adding exponents

Contrasting Relationships
Level 3 Variation: From difference within to difference between

Difference Within (Level 2)

a) $3^4 \times 3^3$
    $= (3 \times 3 \times 3 \times 3) \times (3 \times 3 \times 3)$
    $= 3 \times 3 \times 3 \times 3 \quad \times \quad 3 \times 3 \times 3$
    $= 3^7$

b) $8^4 \times 8^3$
    $= 8 \times 8 \times 8 \times 8 \quad \times \quad 8 \times 8 \times 8$
    $= 8^7$

c) $8^4 \times 8^5$
    $= 8 \times 8 \times 8 \times 8 \quad \times \quad 8 \times 8 \times 8 \times 8 \times 8$
    $= 8^9$

d) $8^4 \times 9^5$
    $= 8 \times 8 \times 8 \times 8 \quad \times \quad 9 \times 9 \times 9 \times 9 \times 9$
    $= 8^4 \times 9^5$ (can't be reduced)

Difference Between (Level 3)

a) $3^2 \times 3^6$
    $= 3 \times 3 \quad \times \quad 3 \times 3 \times 3 \times 3 \times 3 \times 3$
    $= 3^{2+6} \;=\; 3^8$

b) $(3^2)^6$
    $= (3 \times 3)^6$
    $= 3 \times 3 \quad \times \quad 3 \times 3 \quad \times \quad 3 \times 3 \quad \times \quad 3 \times 3 \quad \times \quad 3 \times 3 \quad \times \quad 3 \times 3$
    $= 3^{6 \times 2} \;=\; 3^{12}$

c) $3^6 \div 3^2 = 3^4$
    $= \dfrac{3 \times 3 \quad \times \quad 3 \times 3 \quad \times \quad 3 \times 3}{3 \times 3} \;=\; 3^{6-2} \;=\; 3^4$

d) $3^6 \div 3^4 = 3^2$
    $= \dfrac{3 \times 3 \quad \times \quad 3 \times 3 \quad \times \quad 3 \times 3}{3 \times 3 \times 3 \times 3} \;=\; 3^{6-4} \;=\; 3^2$

Fig. 12.11 Level 3 variation: Contrasting exponent laws

and then identifying what they have in common, here the focus is on impact of change.

This pattern or variation may still be seen in terms of Lai and Murray's PV1 (changing features), but there are also elements of PV2, or changing strategies. In our case, this doesn't show up in the sense of multiple strategies to solve the same problem but through the manner in which both Level 3 and PV2 prompt to contrasting relationships. In fact, Level 3 may be seen as a version of PV2 that varies the relationship against a constant background instead of varying the representation or context against a constant relationship. This isn't to say that there isn't value in comparing carefully selected multiple strategies (Durkin et al. 2017), but for reasons that we explain a bit later, doing so is better described by Level 5 (integration) in our scheme.

Level 4: Abstraction

At Level 4, the relationships among relationships themselves become the focus of attention through contrast with other situations that partially share explanatory structure; meaning may move in both directions, but it's helpful when at least one situation is clearly understood. The example we offer in Fig. 12.12 may seem like an application rather than an abstraction, but the point is that it offers a space where the structure of relevant relationships may be contrasted in ways that allow transfer of meaning.

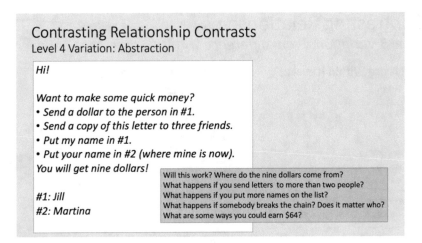

Fig. 12.12 Level 4 variation—Contrasting relationship contrasts: Making sense of exponential structure

The variables in the chain letter (i.e., number of people you send the letter to, number of people on the list, number of people who break the chain) could be seen in terms of variable bases and exponents, but recognizing this as a possibility and considering the appropriateness of transfer is key to making sense of the task. Consideration of combination locks with variable numbers of both numbers (or other symbols) on the lock and numbers (or other symbols) in the combination would serve a similar purpose. The variations described here are partially consistent with Lai and Murray's PV3, which involves "multiple applications of a method by applying the same method to a group of similar problems" (p. 8). Again, however, our emphasis is not on whether the task offers an *application* but on whether the similarity between problems affords transfer of explanatory structure (which is consistent with the examples offered by Lai and Murray).

Lest it seem that we are contradicting ourselves in recommending another strategy that explicitly focuses on identifying similarity, note that mapping an analogy differs from induction in essential ways. *Induction* inappropriately focuses on perception of similarity in that it requires learners to find commonalities among features they have not yet discerned. Earlier, we highlighted how generalization focuses on similarity between previously discerned *features*; here, we highlight how *abstraction* focuses on similarity between previously discerned *relationships*.

Level 5: Integration

Level 5 focuses on the use of tasks that require the integration of seemingly unrelated concepts that become enmeshed in the "same story" and thereby assume relevance to one another. In doing so, Level 5 incorporates all levels of the Level 1–4 hierarchy without adding another layer of differences among differences; it does so by bringing together ideas previously formed through their own progressions. We distinguish integration from abstraction in that where abstraction focuses on the transfer of meaning between two at least partially familiar situations, integration requires the combining of familiar ideas to solve a problem. Integration, then, does not fit into the same hierarchical structure described in Levels 1–4.

The rope metaphor we used to describe the importance of integrating prior knowledge that is well-understood—of taking care not to ask students not to braid the strands of a rope at the same time that they're asked to braid those strands into a rope—is key to Level 5. If raveled appropriately, the "How Many Factors" task we introduced earlier (Fig. 12.5) may be seen as a Level 5 task: It *requires* and therefore *integrates* ideas developed in multiple Level 1–4 progressions, including the one pertaining to exponents developed here. Thus, it might be considered a Level 5 task in a variety of progressions, depending on the order in which topics were introduced. The key point is that each of the components *has* been previously developed before we ask learners to integrate them.

Again, it may seem that integration is mere application. Both abstraction and integration, may (but need not) overlap with applications, but the notion of application doesn't seem to be a particularly useful distinction when considering how a problem or task set prompts to meaning. Similarly, the use of multiple strategies to solve a problem (and considering how they're related) may focus either on varying relationships (Level 3) or on integrating diverse ideas, which we argue is an important distinction.

A Brief Note on Problem Solving

Various notions of problem solving assume relevance in different places within the hierarchy we've developed here, particularly those that focus on general heuristics and those that emphasize transfer to novel situations. A clear focus on the ongoing structuring of knowledge engages learners directly with the sorts of ideas typically highlighted in lists of problem solving strategies, and the dynamic structures thus developed lend themselves metaphoric transfer. Further, Levels 1–3 focus on what is sometimes referred to as "working systematically." Here, working systematically is focused neither on procedural steps with clear worked examples nor "rich tasks" with a focus on "mathematical process."

Teaching with variation *models* working systematically (with structured variation) and weaves together mathematical ideas that *support* working systematically (e.g., identifying combinations, graphical representation, algebraic representation). In other words, ideas are raveled into co-amplifying ideas that serve *each other*. As learners become more familiar with using structured variation and more aware of dimensions that are vary-able, they can take greater responsibility for creating the variations that prompt to new meaning. Working systematically typically gets short-changed in one of two ways: (1) When it's seen as *only* a process for approaching other content, it doesn't get adequately raveled in its own right and (2) when it's seen *only* as an isolated body of content, the important role it *does* play in making sense of other content gets overlooked. Both matter—and both are borne of the same artificial process-content dichotomy.

Levels 4 and 5 address problem solving as application, but application is divided into two categories that differ in terms of the ways they support structuring meaning: Level 4 is about recognizing structural similarity among diverse problems, while Level 5 is requires decomposing complex problems into manageable sub-problems (here "manageable" includes requisite prior knowledge). In other words, Levels 4 and 5 draw from and elaborate the structured knowledge developed in Levels 1–3. Various conceptions of modeling might similarly be categorized according to their role in structuring meaning.

While various discourses on mathematical problem solving (English and Gainsburg 2016; Liljedahl et al. 2016) acknowledge the role of prior knowledge, they focus less on the long-term structuring of that knowledge than on the immediate actions (or non-actions) taken in the hopes of calling forth or generating a fruitful combination of ideas relevant to a particular problem situation. While problem-solving heuristics may support mathematical creativity, discovery, and invention, they can easily lend themselves to an air of mystique that perpetuates the myth that mathematics is primarily the realm of those with a particular type of ability or even genius. We hope that our emphasis on long-term structuring helps create rich ground from which all learners may share in the powerful a-has that accompany moments of illumination and insight.

Summary

Most reports of variation pedagogy focus on a short-term ravel, likely with the assumption that what is well-integrated locally will also be well-integrated on a broader scale. As we observe variation pedagogy being taken up outside of China, this does not seem to be a well-grounded assumption. Collections of isolated lessons, even when well-varied internally (i.e., even when the focus is on mathematical structure and not merely on procedures), do little to prompt to the broader integrated structure of mathematics, and most curricula and resources are grossly insufficient in supporting such coherence (Bajaj 2013).

It is a monumental undertaking to build an effective sequence of effectively varied lessons. Even when thoughtfully designed, such lessons inevitably change when they meet students—not just to suit the idiosyncrasies of individual students, but in ways that gradually become more consistent with the phenomenographic space of possibility defined by the affordances of human perception. It is not reasonable to expect individual teachers to take on the task of re-inventing and refining effective long-term mathematical sequences. Even when teachers have sufficient pedagogical content knowledge relevant to their grade level, there is much that can be offered in the form of a well-raveled resource that takes into account both the structure of mathematics, common patterns of interaction between learners and mathematics, and long-term coherence between grades. With this in place, teachers are simultaneously supported in making sense of the long-term ravel and freed to attend to the fine-grained variations relevant to their moment-by-moment interactions with students.

Conclusion

As both teachers and researchers, we must confess to a frustration with the field of education. At times it feels as though there is no other domain of human engagement that is more resistant to well-structured theory (e.g., exposing the metaphoric substrate of entrenched activity) and validated evidence (e.g., year-over-year improvements in learner engagement and understanding). When presented with such ideas and evidence, more often than not, the system finds ways to reject or minimize it—often by characterizing a new insight as reflective of the other "camp" in whatever skirmish is happening at the moment.

It's thus that we have experienced criticisms and rejections from teachers and policy makers positioned at both poles in the Math Wars. For instance, Traditionalists balk at the assertions that *all* learners can become adept at mathematics, that gaps in understanding are attributable to missed noticings, and that perceived differences in learner ability have more to do with flawed pedagogy than flawed learners. On the other side, Reformists have a strong tendency to see a well-deconstructed concept in terms of the much-hated step-based approach to teaching rather than an equity-informed noticing approach. Absences are another favorite focus of their criticisms. Limited group work, few open-ended problems, and no heed to personalized strategies—notwithstanding that the evidence supporting such emphases is dubious at best—are lobbed as reasons to reject the model entirely. We actually take these rejections by both staunch Traditionalists and staunch Reformists as positive indicators, emboldened by Dewey's (1990: 9) observation regarding oppositional thinking, noted earlier: "We don't solve them: we get over them."

Part of that "getting over" is hinged to rethinking the relationship between teachers and resources. As we hope is evident in the preceding discussion, a well-structured inquiry involves high levels of knowledge and extensive effort. Each of our lessons has pulled in the expertise of mathematicians, logicians, teachers, and educational researchers. Flatly stated, there is no way that solitary teachers in isolated classrooms might be reasonably expected to design such lessons on their own.

It is thus that we frame fitted-to-the-Information-Age approach to mathematics teaching in terms of a partnership in which each aspect is associated with differentiated obligations. Principal responsibility for the Ravel—that is, discerning the critical discernments involved in a concept, appreciating their relationships to one another, and so on—sits with the resource. Responsibility for the Prompt is more shared, with the resource providing suggestions and that teacher selecting and adapting those suggestions according to knowledge of experiences, established competencies, and interests of those present. The contingencies associated with Interpreting means that that element is almost entirely the responsibility of the teacher, and Deciding what to do next is a shared responsibility that sits across the teacher's knowledge of what's happening and the resource's advice on what might happen next. Conceived as a partnership, the RaPID model is neither a step-following (Traditionalist) script nor an open-ended (Reformist) exploration.

To state this point more emphatically, we see the next moment in the necessary evolution of school mathematics in the Information Age to be about a much-expanded and formalized partnership between teachers and resources, each having obligations to the other.

That suggestion is heresy within much of the current educational establishment. It strikes against two principles that are held by Traditionalist and Reformist alike: firstly, a conviction on the sanctity of teacher autonomy and, secondly, a belief that the best response to learner difference is differentiated experience. We question those ideals. In an era of massive connectivity (in which there can be genuine, mutually beneficial influences between teachers and resource developers) and better understandings of knowledge and cognition (that point to flawed assumptions in differentiated models of instruction), new possibilities for school mathematics are not just afforded, they are required.

References

Bajaj, V. (2013). Q and A with Liping Ma. *The New York Times*. From https://www.nytimes.com/2013/12/18/opinion/q-a-with-liping-ma.html

Bateson, G. (2002). *Mind and nature: A necessary unity.* Cresskill: Hampton Press. (Original work published 1979).

Chazan, D. and Ball, D. (1999). Beyond being told not to tell. *For the Learning of Mathematics* 19: 2–10.

Davis, B. (2015). Exponentiation: A new basic? *Mathematics Teaching in the Middle School* 21: 34–41.

Dewey, J. (1910). *The influence of Darwin on philosophy and other essays.* New York: Henry Holt.

Dienes, Z. (1960). *Building up mathematics.* New York: Hutchinson.

Durkin, K., Star, J., and Rittle-Johnson, B. (2017). Using comparison of multiple strategies in the mathematics classroom: lessons learned and next steps. *ZDM Mathematics* 49: 585–597.

English L. D. and Gainsburg, J, (2016). Problem solving in a 21st century mathematics curriculum. In: L. D. English and D. Kirshner (eds.), *Handbook of international research in mathematics education* (3rd edition), 313–335. New York: Routledge.

Foote, R. (2007) Mathematics and complex systems. *Science* 318: 410–412.

Gu, F., Huang, R., and Gu, L. (2017). Theory and development of teaching through variation in mathematics in China. In: R. Huang and Y. Li (eds.), *Teaching and learning mathematics through variation: Confucian heritage meets Western theories,* 13–41. Boston: Sense Publishers.

Gu, L., Huang, R., and Marton, F. (2004). Teaching with variation: A Chinese way of promoting effective mathematics learning. In: L. Fan, N.-Y. Wong, J. Cai, and S. Li (eds.), *How Chinese learn mathematics: Perspectives from insiders,* 309–347. Hackensack: World Scientific.

Hewitt, D. (1992). Train-spotters' paradise. *Mathematics Teaching* 140: 6–8.

Hewitt, D. (1999). Arbitrary and necessary: Part 1: A way of viewing the mathematics curriculum. *For the Learning of Mathematics* 19: 2–51.

Hewitt, D. (2001a). Arbitrary and necessary: Part 2: Assisting memory. *For the Learning of Mathematics* 21: 44–51.

Hewitt, D. (2001b). Arbitrary and necessary: Part 3: Educating awareness. *For the Learning of Mathematics* 21: 37–49.

Huang, R., Barlow, A., and Prince, K. (2016). The same tasks, different learning opportunities: An analysis of two exemplary lessons in China and the U.S. from a perspective of variation. *The Journal of Mathematical Behavior* 41: 141–158.

Huang, R. and Leung, K. S. F. (2004). Cracking the paradox of Chinese learners. In: L. Fan, N.-Y. Wong, J. Cai, and S. Li (eds.), *How Chinese learn mathematics: Perspectives from insiders*, 348–381. Hackensack: World Scientific.

Jianhua, L. I. (2004). Thorough understanding of the textbook—A significant feature of Chinese teacher manuals. In: L. Fan, N.-Y. Wong, J. Cai, and S. Li (eds.), *How Chinese learn mathematics: Perspectives from insiders*, 262–281. Hackensack: World Scientific.

Kullberg, A., Runesson Kempe, U., and Marton, F. (2017). What is made possible to learn with the variation theory of learning in teaching mathematics? *ZDM Mathematics Education* 49: 559–569.

Lakoff, G., and Núñez, R. (2000). *Where mathematics comes from*. New York: Basic Books.

Lai, M. Y. and Murray, S. (2014). Teaching with procedural variation: A Chinese way of promoting understanding of mathematics. *International Journal for Mathematics Teaching and Learning*. from http://www.cimt.org.uk/journal/lai.pdf.

Liljedahl, P., Santos-Trigo M., Malaspina, U., and Bruder, R. (2016). *Problem solving in mathematics education*. Chaim: Springer Nature.

Ma, L. (1999). *Knowing and teaching elementary mathematics*. Mahwah: Lawrence Erlbaum.

Marton, F. and Booth, S. (1997). *Learning and awareness*. Mahwah: Lawrence Erlbaum.

Marton, F. and Tsui, A. B. M. (2004). *Classroom discourse and the space of learning*. Mahwah: Lawrence Erlbaum.

Marton, F. (2015). *Necessary conditions of learning*. New York: Routledge.

Mason, J. (2010). Effective questioning and responding in the mathematics classroom. from http://mcs.open.ac.uk/jhm3/Selected%20Publications/Effective%20Questioning%20and%20 Responding.pdf

Mason, J. and Pimm, D. (1984). Generic examples: Seeing the general in the particular. *Educational Studies in Mathematics* 15: 277–289.

Mason, J., Burton, L., and Stacey, K. (2010). *Thinking mathematically* (2nd edition, New York: Prentice Hall.

Math Minds. (2020). *Math minds online course*. https://www.structuringinquiry.com.

Preciado-Babb, A. P., Metz, M., Davis, B., and Sabbaghan, S. (2020). Transcending contemporary obsessions: The development of a model for teacher professional development. In: S. Linares and O. Chapman (eds.), *International handbook of mathematics teacher education*. 361–390. Boston: Sense Publishers.

Ong, W. J. (1982). *Orality and literacy: The technologizing of the word*. New York: Routledge.

Organization for Economic Cooperation and Development. (2010). *Recognising non-formal and informal learning: Outcomes, policies and practices.* http://www.oecd.org/education/innovation-education/recognisingnon-formalandinformallearningoutcomespoliciesandpractices.htm

Pang, M.F., Marton, F., Bao, J., and Ki, W.W. (2016). Teaching to add three-digit numbers in Hong Kong and Shanghai: Illustration of differences in the systematic use of variation and invariance. *ZDM Mathematics Education* 48: 455–470.

Runesson, U. (2005). Beyond discourse and interaction. Variation: A critical aspect for teaching and learning mathematics. *Cambridge Journal of Education* 35: 69–87.

Schoenfeld, A. H. (2004). The math wars. *Educational Policy* 18: 253–286.

Sfard, A. (1998). On two metaphors for learning and the dangers of choosing just one. *Educational Researcher* 27: 4–13.

Skemp, R. (1976). Relational understanding and instrumental understanding. *Mathematics Teaching*, 77: 20–26.

Sun, X. (2011). "Variation problems" and their roles in the topic of fraction division in Chinese mathematics textbook examples. *Educational Studies in Mathematics* 76: 65–85.

Watson, A. (2017). Pedagogy of variations: Synthesis of various notions of variation pedagogy. In: R. Huang and Y. Li (eds.), *Teaching and learning mathematics through variation: Confucian heritage meets Western theories,* 85–103. Boston: Sense Publishers.

Watson, A. and Mason, J. (2005). *Mathematics as a constructive activity: Learners generating examples.* Mahwah: Lawrence Erlbaum.

Watson, A. and Mason, J. (2006). Seeing an exercise as a single mathematical object: Using variation to structure sense-making. *Mathematical Thinking and Learning* 8: 91–111.

Wong, N. Y. (2008). Confucian heritage culture learner's phenomenon: from "exploring the middle zone" to "constructing a bridge." *ZDM Mathematics Education* 40: 973–981.

Chapter 13
If One Can Read and Write Then One Can Also Do Mathematics

Robert K. Logan

Introduction

There are those that claim that they are good at reading and writing but they are not as good when it comes to mathematics. Actually the 3 R's of reading, 'riting and 'rithmetic are more closely related than most people think. I will show in this essay that the origin of verbal language and mathematics depend on each other. In particular the mathematical skill associated with set theory is what gave rise to the origin of verbal language and verbal language allowed mathematics to evolve from a primitive set theory to arithmetic and all the beautiful structures of mathematics afterwards. So, if one is good at talking, reading and writing they have no excuse for doing poorly at mathematics especially arithmetic. It is important that we dispel this erroneous notion that one can be good with verbal language and be a disaster with numbers and mathematics. The manipulation of information which requires mathematical skills is as important in the information age and as being literate was in the age of the written word. Mathematics, as we will argue, is a language and today mathematical literacy is just as important as verbal literacy. So, let's get started and examine the topology of mathematics in the mind and its interaction with verbal and written language from which it emerged and examine its connection with science, computing, the Internet and the World Wide Web, the languages of the Information Age. In fact, we will explore the notion that mathematics is part of an evolutionary chain of six languages. One does not ordinarily think of computing and the Internet as languages but I will demonstrate that speech, writing, mathematics, science, computing and the Internet form an evolutionary chain of six languages.

R. K. Logan (✉)
Physics Department, Emeritus. Studies Media Ecology, Media, and Information Theory,
University of Toronto, Toronto, ON, Canada
e-mail: logan@physics.utoronto.ca

© Springer Nature Switzerland AG 2020   219
S. A. Costa et al. (eds.), *Mathematics (Education) in the Information Age*,
Mathematics in Mind, https://doi.org/10.1007/978-3-030-59177-9_13

What Is a Language and What Is Its Dynamic Nature?

Language is not the passive container or medium of human thought whose only function is to transmit and communicate our ideas and sentiments from one person to another. Language is a "living vortices of power" (Innis 1972, v. McLuhan's foreword), which shapes and transforms our thinking. Language is both a system of communications and an informatics tool. Without verbal language our mental life would be reduced to feelings, emotions, and the processing of our perceptions as is the case with all other forms of life. Verbal language makes conceptualization, abstraction and reflection possible. Our ability to use verbal language is what differentiates us from the rest of the animal kingdom. Other animals are capable of communicating with each other but the range of what they can express is limited to a small number (less than 50) of signals. Human verbal language, on the other hand, is generative so there are an infinite number of possible messages or meaning that we are capable of composing and communicating. Abstract thinking and language cannot be separated.

The Origins of Verbal Language

Only humans are capable of verbal language, abstract thought, mathematics and conceptualization. It is believed that first came verbal language and then mathematical thought. But I suggest otherwise. The origin of verbal language, the origin of the mind and the origin of mathematic thinking all happened at approximately the same time and that these three elements are basically interlinked. Human verbal language was as much a product of mathematical thinking as mathematics was a product of verbal language.

The human mind is a product of the brain and verbal language as was argued in *The Extended Mind: The Emergence of Language, the Human Mind and Culture* (Logan 2007), but verbal language as we have argued was dependent on the ability of humans to think in terms of sets employing a primitive form of set theory. Before verbal language, we lived in a world of percepts. Our communication was mimetic consisting of hand signals, facial gestures, body language and non-verbal prosody or tones such as grunts and whines. We could only communicate about the here and now.

Conceptual thinking only became possible with verbal language and our first concepts were our first words. These words acting as concepts linked to and representing all the percepts associated with those words. For example, the word water represents the concept of water and instantaneously triggers all of the mind's direct experiences and perceptions of water such as the water we drink, the water we cook with, the water we wash with, the water that falls as rain or melts from snow and the water that is found in rivers, ponds, lakes, and oceans.

The word "water" acting as a concept and an attractor not only brings to mind all "water" transactions but it also provides a name or a handle for the concept of water, which makes it easier to access memories of water and share them with others or make plans about the use of water. Words representing concepts speed up reaction time and, hence, confer a selection advantage for their users. And at the same time those languages and those words within a language, which most easily capture memories enjoys a selective advantage over alternative forms of communication.

The skill that made language possible and allowed a word acting as a concept to represent all of the percepts associated with that word was the mathematical ability to create sets, the set of all the percepts associated with that word. We suggest that the brain before verbal language was merely a percept processor and that afterwards it was able to conceptualize, i.e. operate with concepts. Each concept linked all the percepts associated with that concept. We conclude that the human mind naturally makes associations, creates categories or sets and hence has the natural mathematical structure of set theory.

We further suggest that verbal language emerged as a primitive form of set theory in that a set of percepts that are associated with each other or are similar are linked together with a word acting as a concept that unites all the members of that set. In a certain sense the primitive form of set theory we just described seems to be a pre-condition for the emergence of verbal language. It is not possible to determine the causal linkage between the primitive form of set theory and verbal language. It is not that set theory caused verbal language to emerge or that language allowed set theory to emerge. Rather we would claim, invoking complexity theory and emergent dynamics, that mathematical set theory and verbal language self-organized into an emergent supervenient system.

In our model, the emergence of set theory preceded the emergence of enumeration as enumeration requires verbal language. There are two types of numbers, concrete numbers and abstract numbers. A pair of shoes or a yoke of oxen are concrete numbers. Concrete numbers have meaning only as units of the commodity they are designating and enumerating. Concrete numbers cannot represent abstract numbers like one, two or three. The number 'two' is abstract as it can apply to any set of two objects. We would surmise that at some point in the evolution of language one particular concrete number came to represent an abstract number. We can only guess as to how this happened as there is no record of how verbal language evolved from its origins, but certainly it is the case that numbers in the form of numerals like one, two and three are basically concepts represented by words. A hint of how the evolution from concrete numbers to abstract numbers might have occurred in verbal language might be ascertained by studying how notated concrete numbers evolved into notated abstract numbers which we will consider shortly.

The model that we have proposed of how verbal language and mathematical thinking co-emerged is an abduction or a just so story. It is a hypothesis but it cannot rise to the level of a scientific hypothesis because it cannot be falsified as the emergence of verbal language happened long before any scientific observations could be made. But in my mind even though it is a just so story it might just be true.

Mathematics in the Mind Leads to Writing in Sumer and Writing Leads to the Further Development of Mathematical Thinking

Not only did mathematical thinking lead to verbal language but it also gave rise to written language through the development of mathematical notation. The very first notation for recording quantities were tally sticks. The tally stick, however, gave no indication of what was being tallied but they were the first forms of notated concrete numbers.

The next step in the evolution of numerical notation were three-dimensional clay accounting tokens that archeologist Denise Schmandt-Besserat discovered in the Fertile Crescent of Mesopotamia between the Tigris and Euphrates rivers that were used from 8000 BCE to about 3000 BCE. These tokens had different shapes that corresponded to the things that they were enumerating which were agricultural commodities and hence represent a step forward from tally sticks in the evolution of numerical notation. For example, one token shape represented a large measure of wheat, another token shape a small measure of wheat and a third token shape a jar of oil. The tokens were used as receipts for tributes paid by farmers to the priest-accountants as a form of taxation. These tributes were redistributed to the irrigation workers who provided the water that was essential for the farmers to grow their crops. These clay tokens did not represent abstract numbers like 1, 2 or 3 but they were concrete numbers. Three 'jar of oil' tokens did not represent the abstract number three but rather represented 3 jars of oil and two 'large measure of wheat' tokens did not represent the abstract number two but rather represented 2 large measures of wheat. These tokens representing concrete numbers, however, evolved into abstract numbers as we shall now relate.

The clay tokens were stored in clay envelopes starting around 3200 BCE to make sure that the tokens would not get lost. Then because it was a nuisance to have to break open the clay envelopes to see what was inside and reseal the token in a new clay envelope it was decided to press the tokens into the surface of the clay envelopes so one could determine what was inside without having to break open the envelope. It then occurred to a smart accountant that there was no need for the envelopes. One could merely press the tokens into a clay tablet and then one would have a permanent record of the accounting. The token impressions on the clay tablets still represented concrete numbers but this practice would eventually lead to abstract numbers.

What led to abstract numbers or numerals was the combination of the clay tablets with their concrete numbers and an increase in the commerce in Sumer in Mesopotamia about 3000 BCE. As commercial transactions began to incorporate large quantities of commodities the system of pressing clay tokens many into a tablet became cumbersome and unmanageable. To deal with this challenge a new system of accounting emerged making use of abstract numbers.

The large and small measure of wheat were used to represent the abstract numbers "10" and "1" respectively. These tokens were still pressed into clay tablets but the tokens for the other commodities were no longer pressed into the clay, but rather the shapes that these tokens would make if pressed into the clay were drawn with a stylus. As a result, two classes of signs emerged. The impressed token shape signs for the large and small measure of wheat came to represent the numerals 10 and 1 respectively. The second class of signs, etched signs representing the commodities and hence represented words. Once the idea of using signs to represent words came into practice it was soon realized that all spoken words could be represented by etched signs and hence the idea of writing emerged from the accounting token system. The Sumerians quickly realized that they could represent spoken words other than agricultural commodities with etched signs and so writing emerged to represent spoken language and also abstract numbers. Writing was invented by accountants and not by writers, but writers harnessed the accountant's idea of creating written symbols for spoken words. And mathematicians harnessed the accountant's idea of notating abstract numbers to develop a notation for mathematics that led to more sophisticated mathematical thinking.

The idea of writing and mathematical notation spread from Sumer throughout the Eastern Hemisphere. First by the Egyptians and the Semitic people in the Levant and then the Greeks and from them all over the Middle East and Europe. The idea of writing and mathematical notation also spread to China over the trade routes between the Middle East and the Orient. The independent invention of writing in the Western Hemisphere began in Mesoamerica beginning with the Zapotec writing system that has not yet been fully deciphered. We therefore cannot find a link between math and writing for the Mesoamerican writing systems as we do not know how that system emerged, but the Mesoamerican writing system also included a notation for abstract numbers.

Aztec numerals used the following symbols: a dot for 1; a bar for 5; and this shape ⬯ for 20 or sometimes for zero. They also deployed a place number system for larger numbers based on 20, $400 = 20^2$ and $8000 = 20^3$. One of the chief uses of the Aztec writing was to keep track of the Mesoamerican calendar providing a possible hint of a connection between math and writing.

With a written notation for both words and mathematical notation not only was communication enhanced but mathematical thinking became more sophisticated. De Cruz and De Smedt (2013: 3) argue that

> mathematical symbols are not only used to express mathematical concepts—they are constitutive of the mathematical concepts themselves. Mathematical symbols are epistemic actions, because they enable us to represent concepts that are literally unthinkable with our bare brain [signaling] an intimate relationship between mathematical symbols and mathematical cognition.

Thus, mathematical thinking gave rise to mathematical notation and writing which in turn led to the further development of mathematical thinking. Having demonstrated how speech, writing and mathematics are interlinked we now turn to show how they are actually part of a larger evolutionary chain of interlinked languages composed of speech, writing, mathematics, science, computing and the Internet.

An Evolutionary Chain of Six Languages: Speech, Writing, Mathematics, Science, Computing and the Internet

Marshall McLuhan (1964: 8) in his ground-breaking study of media titled *Understanding media* noted that

> the 'content' of any medium is always another medium. The content of writing is speech, just as the written word is the content of print, and print is the content of the telegraph.

The content of the Internet is computing and the content of computing is science and the content of science is mathematics and writing and the content of mathematics and writing is speech. When a medium first appears, it uses the content of another medium exclusively for its content until its users have learned to exploit the new medium to develop new forms of expression. We saw that writing and mathematical notation made use of spoken language for its content. Science makes use of the spoken word, writing and mathematics. The practice of computing makes use of the skills of spoken language, writing, mathematics, and science. And finally, the development of the Internet, the language of the Information Age, required all five of the languages that preceded it, namely speech, writing, mathematics, science and computing.

The Dynamic Nature of Language

Language is not the passive container or medium of human thought whose only function is to transmit and communicate our ideas and sentiments from one person to another. As McLuhan (1972) noted language is a "living vortices of power," which shapes and transforms our thinking. Language is both an informatics tool and a system of communications. One cannot separate language from thinking. Without verbal language our mental life would be reduced to the processing of our perceptions, feelings and emotions. Verbal language allows conceptualization, abstraction and reflection to take place. Our ability to use verbal language is what differentiates us from the rest of the animal kingdom. Other animals are capable of communicating with each other but the range of what they can express is limited to small number of signals. They are unable to conceptualize. They have no sense of the past and the future. They live in the perpetual present. Human verbal language allows us to think about abstractions, to plan for the future and reminisce about the past. Language is generative so that we are capable of creating an infinite number of possible messages or ideas.

Language is like an organism that grows and evolves, with both its semantics and syntax becoming ever more complex. Language began as verbal or spoken language and evolved into written language, mathematics, science, computing, and the Internet (Logan 2004). And each of these six languages have undergone their own individual evolutions. Each of these six languages is like a symbiotic organism that depends on its human hosts for its sustenance.

Each mode of language incorporates the features of the previous modes. The Internet incorporates all of the features of computing which in turn incorporates the features of all the previous modes: speech, writing, mathematics, and science. Science incorporates speech, writing, and mathematics. Writing and mathematics arose at the same moment in history. They therefore only incorporated the features of speech, albeit different ones.

In addition to each new language incorporating the features of the languages that preceded it, it is also the case that each new language changed the languages that preceded it. Writing changed the way spoken language was used. The first uses of spoken language were for the purpose of communication. However, it was discovered that verbal language could also be used to record and store information through the mechanisms of poetry and song. This discovery gave rise to the oral tradition among preliterate people whereby vital information necessary for their survival, their identity, and their sense of history was recorded or stored in their tales, epics, and legends that were in the form of poetry that facilitated their being memorized. Oral language was used as a device for processing, storing, retrieving, and organizing information. The preservation or storage of information and knowledge in preliterate societies was achieved through the memorization of folktales or myths. These tales or legends were not merely entertaining stories told in an impromptu manner. They were, in fact, very carefully organized to provide listeners with the basic information required in their society. In his Preface to Plato, Eric Havelock refers to the oral storyteller as a "tribal encyclopedia." With the advent of writing oral poetry was no longer required to store information and as a result poetry became an art form rather than a way to store information.

Writing and mathematical notation emerged at the same time in Sumer in the transition from three-dimensional accounting tokens to a written notation for numbers and words on clay tablets. Writing changed the nature of mathematics as formal proofs of geometric relationships emerged in ancient Greece as well as deductive logic. Mathematics in the age of alphabetic literacy changed just as mathematics is changing in the age of information. Alphabetic literacy is not only associated with deductive logic but McLuhan and Logan (1977) showed an interesting connection of alphabetic literacy with codified law, monotheism, abstract science and deductive logic.

> Western thought patterns are highly abstract, compared with Eastern. There developed in the West, and only in the West, a group of innovations that constitute the basis of Western thought. These include (in addition to the alphabet) codified law, monotheism, abstract science, formal logic, and individualism. All of these innovations, including the alphabet, arose within the very narrow geographic zone between the Tigris-Euphrates river system and the Aegean Sea, and within the very narrow time frame between 2000 B.C. and 500 B.C. We do not consider this to be an accident. While not suggesting a direct causal connection between the alphabet and the other innovations, we would claim, however, that the phonetic alphabet played a particularly dynamic role within this constellation of events and provided the *ground* or framework for the mutual development of these innovations.

This observation of the effects of phonetic writing supports the notion that mathematics will change in the age of information, i.e. the age of computers and the Internet. Mathematics is, in fact, changing as is reported in the Wikipedia article Computer-assisted proof (https://en.wikipedia.org/wiki/Computer-assisted_proof).

Another example of how a new language changes a language that preceded it is the way the emergence of abstract science affected mathematics with scientists developing new kinds of mathematics to describe nature. Descartes's analytic geometry and Newton's differential calculus being two examples of this among many other examples. The effects of computing on science are far too many to document. Suffice it to say that almost every contemporary scientific discovery involves the use of the computer. The language of the Internet and the dialect of the World Wide Web are essential for scientific activity. In fact, the Web was developed at the elementary particle accelerator at CERN by Tim Berners-Lee (1999) to facilitate the communications of the international research teams that carried out their research at CERN.

Conclusion

If one is to operate successfully in the Information Age one must be fluent with all six languages especially mathematics as it is the base for computing and using the Internet. Mathematics in the age of information and the age of computers, the Internet and the World Wide Web will continue to evolve. It is also likely that as new languages emerge, such is the case with robotics and artificial intelligence that they will inspire even more new mathematics. I cannot predict was these will be but I am confident that the complexity of mathematics in the Age of Information will continue to grow. QED!

References

Berners-Lee, T. (1999). *Weaving the web*. San Francisco: Harper.
De Cruz, H. and De Smedt, J. (2013). Mathematical symbols as epistemic actions. *Synthèse* 190: 3–19.
Havelock, E. (1963). *Preface to Plato*. Cambridge: Harvard Univ. Press
Logan, R. K. (2004). *The sixth language: Learning a living in the Internet Age*. Caldwell: Blackburn Press.
McLuhan, M (1964). *Understanding media: Extensions of man*. New York: McGraw Hill.
McLuhan, Marshall, and Robert.K. Logan. 1977. Alphabet: Mother of Invention. Etcetera 34: 373-83.
McLuhan, M. (1972). McLuhan's foreword to the 1972 edition of Harold Innis' *Empire and communication*. Toronto: Dundurn Press.
McLuhan, M. and Logan, R. K. (1977). Alphabet: Mother of invention. *Etcetera* 34:373–383.
Schmandt-Besserat, D. (1978). The earliest precursor of writing. *Scientific American* 238: 50–58.
Schmandt-Besserat, D. (1986). The origins of writing: An archaeologist's perspective. *Written Communication* 3: 31–45.

Printed in the United States
by Baker & Taylor Publisher Services